D0023071

Chemical Instrumentation

Richard P. Wayne

The Physical Chemistry Laboratory and Christ Church, University of Oxford

OXFORD   NEW YORK   TOKYO
OXFORD UNIVERSITY PRESS
1994

Oxford University Press, Walton Street, Oxford OX2 6DP

Oxford New York
Athens Auckland Bangkok Bombay
Calcutta Cape Town Dar es Salamm Delhi
Florence Hong Kong Istanbul Karachi
Kuala Lumpur Madras Madrid Melbourne
Mexico City Nairobi Paris Singapore
Taipei Tokyo Toronto

and associated companies in
Berlin Ibadan

Oxford is a trade mark of Oxford University Press

Published in the United States
by Oxford University Press Inc., New York

© Richard P. Wayne, 1994

All rights reserved. No part of this publication may be
reproduced, stored in a retrieval system, or transmitted, in any
form or by any means, without the prior permission in writing of Oxford
University Press. Within the UK, exceptions are allowed in respect of any
fair dealing for the purpose of research or private study, or criticism or
review, as permitted under the Copyright, Designs and Patents Act, 1988, or
in the case of reprographic reproduction in accordance with the terms of
the licences issued by the Copyright Licensing Agency. Enquiries concerning
reproduction outside those terms and in other countries should be sent to
the Rights Department, Oxford University Press, at the address above

This book is sold subject to the condition that it shall not,
by way of trade or otherwise, be lent, re-sold, hired out, or otherwise
circulated without the publisher's prior consent in any form of binding
or cover other than that in which it is published and without a similar
condition including this condition being imposed
on the subsequent purchaser

A catalogue record for this book is available from the British Library

Library of Congress Cataloging in Publication Data
(Data available)

ISBN 0 19 8557973 (Hbk)
ISBN 0 19 8557965 (Pbk)

Typeset by the author
Printed in Great Britain by
The Alden Press, Oxford.

Series Editor's Foreword

Oxford Chemistry Primers are designed to provide clear and concise introductions to a wide range of topics that may be encountered by chemistry students as they progress from the freshman stage through to graduation. The Physical Chemistry series is designed to contain books easily recognised as relating to established fundamental core material that all chemists will need to know, as well as books reflecting new directions and research trends in the subject, thereby anticipating (and perhaps encouraging) the evolution of modern undergraduate courses.

In this Physical Chemistry Primer, Richard Wayne has produced an authoritative, easy-to-read introductory account of Chemical Instrumentation. The Primer introduces essential basic ideas required to understand and exploit the instrumentation encountered by chemistry students in the laboratory at all stages up to and including research level. This Primer will be of lasting and broad value to all students of chemistry (and their mentors).

Richard G. Compton
Physical Chemistry Laboratory, University of Oxford

Preface

The application of electronic instrumentation to chemical experiments has had a dramatic impact on the subject of chemistry itself. Many techniques such as NMR spectroscopy owe their existence to electronics. Others have been made simpler, more reliable, and more precise; electronic data acquisition and signal enhancement techniques permit experiments on time scales and with sensitivities that would otherwise prove impossible. Pervading all these developments in electronics has been the increasing use of computers.

A modern chemist needs to have some idea of how the instrumentation works to exploit its capabilities. This book provides an introduction to the subject by discussing the basic building blocks and techniques of the instrumentation. It is not a book about electronics itself or about detailed circuit design, although the first chapter presents a review of the basic principles. Considerable emphasis is given to the recovery of signals from noise in Chapter 5. Suggestions for further reading are given at the end of the book.

I wish to thank my colleague Pete Biggs for the enormous help he gave me in preparing all the diagrams for the book. His skilled efforts are much appreciated.

Oxford   R.P.W.
January 1994

CENTRAL ARKANSAS LIBRARY SYSTEM
LITTLE ROCK PUBLIC LIBRARY
700 LOUISIANA STREET
LITTLE ROCK, ARKANSAS 72201

Contents

1   Fundamental ideas

Elegant and sophisticated experiments in chemistry are enabling us to understand in increasingly intimate ways the nature of the interactions that lead to chemical change, and to identify chemical species of great complexity present at minute concentrations. Almost all these experiments have depended on the development of sensitive and precise instrumentation that allows measurements to be made and systems to be probed reliably and rapidly. Much of the equipment can be purchased virtually ready for use, although sometimes different elements have to be connected together to satisfy a particular purpose. A 'black-box' approach to chemical instrumentation can be perfectly effective so long as nothing more is asked than the performance that the manufacturers intended. When novel experiments are designed, however, or if apparatus is to be used at the very limits of its potential as is so often the case in creative research, then a different attitude may be needed. It may be necessary, for example, to use the instrumentation in a new way, to make minor modifications to it, or to assemble and connect together units and components to satisfy the special requirements of the work.

This book has been written with the intention of introducing chemists to some of the the building blocks and devices that make up the most important instruments used in industry and research. It should provide a basic understanding of how the instrumentation works and thus help towards using what is available both correctly and effectively; the knowledge may perhaps even provide the first step along the path towards devising novel ways of using existing instrumentation and ultimately of developing entirely new tools by which chemistry may be studied.

Virtually every scientific instrument encodes the information of interest in some form of electrical signal, which is then manipulated as required, and subsequently decoded into a useful form. Many of the advances of instrumentation have therefore been made possible by improvements in electronics; the advent of solid-state devices, first the transistor and then integrated circuits of ever-increasing complexity, has had a major impact on instruments and techniques in all branches of science and technology. Because of the almost universal application of electrically encoded signals, instrumentation and electronics are closely intertwined. However, it is not the intention of this book to dwell unduly on the intricacies of electronics, although some principles must be understood before the operation of the instruments can be explained. The rest of this chapter is aimed, therefore, at those who have forgotten those principles, or were, perhaps, never taught them. Other readers can skip much of it, but might still find the reminders that it contains of value before they tackle the rest of the book.

1.1 Current, potential difference, and power

This section deals with the most fundamental of electrical quantities. A flow of electrons in a conductor constitutes a *current*. If the charge that passes each second is one coulomb, the current is one amp (or ampère). For electrons to flow, one end of the conductor has to be more positive than the other. That is, one end is at a higher potential than the other, and there is a *potential difference* between the ends. It is this potential difference (p.d.) that causes the current to flow. The unit of p.d. is the volt, and p.d. is sometimes referred to in a shorthand way as 'voltage'. We shall adopt this usage throughout the book when it is convenient. Another closely related quantity is the *electromotive force* (e.m.f.), also measured in volts, and which differs from p.d. only in a subtle way that need not concern us here (see p. 23).

If a current flows in a circuit, then charge moves down the potential gradient, and work can be done equal to the charge multiplied by the potential. Since the current is the charge that flows in unit time, the product of current and voltage is the work done in unit time, which is the *power* delivered to the system

$$P = IV \tag{1.1}$$

work = charge × potential

power = work × time
 = charge × potential/time
 = current × potential

where P is the power, I the current and V the voltage. If I is measured in amps and V in volts, then P is in watts.

1.2 Resistance

Any real conductor normally offers a *resistance* to the flow of charge. If it did not, an infinite current would flow for the smallest potential difference. The relationship between the current flowing in the conductor, and the voltage applied across it is embodied in *Ohm's law*, which states that

$$I = V/R \tag{1.2}$$

where I is the current, V the voltage and R the resistance. If I is measured in amps and V in volts, then R is in ohms.

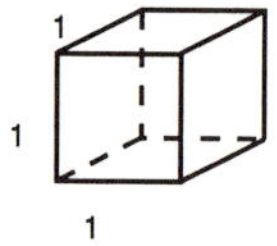

Resistivity and resistors

The resistance of a conductor increases in proportion to its length, and inversely with the cross-sectional area. For a given length and area, different materials exhibit different resistances. The *specific resistivity* is the resistance of a sample of unit length and unit cross-sectional area (cf. Fig. 1.1). Copper and silver are good conductors and have lower specific resistances than many other materials, although aluminium is a relatively good conductor that offers advantages in terms of weight. A *resistor* is a key electronic component that is designed to offer a chosen resistance. Resistors can be made of lengths of wire (usually wound on a former), thin films of metal coated on glass or

Fig. 1.1 The specific resistivity is the resistance between opposite faces of a cube of unit length sides

another insulator, or lengths of rod made of carbon (which is a modest conductor). *Semiconductors* are neither good conductors nor perfect insulators. Some semiconductors, however, possess extraordinary electrical properties that allow the construction of devices such as diodes and transistors (see Section 1.5). Examples include elements such as silicon and germanium, usually 'doped' with an impurity element of a neighbouring group of the periodic table in order to confer the desired properties.

Ohm's law may be used very simply to calculate the power dissipated in a circuit in terms of current and resistance or voltage and resistance. Equations (1.1) and (1.2) need to be combined to give the desired relation, as shown in the margin.

$$P = VI; \quad V = IR; \quad I = V/R$$
$$= I^2 r$$
$$= V^2/R$$

Temperature dependence of resistance

Good metallic conductors usually show a resistance that decreases with decreasing temperature. Indeed, as absolute zero of temperature is approached, the resistance of a conductor may also tend towards zero, and the sample is said to be *superconducting*. Superconducting magnets, cooled by liquid helium to around 4K, find several applications in chemical instrumentation, most notably in nuclear magnetic resonance (n.m.r.) spectrometers. One major advantage of the superconducting magnet is that very large currents (and thus high magnetic fields) can be sustained without appreciable heating. A new class of material that exhibits superconductivity at much higher temperatures has recently been discovered, and offers the prospects of improved performance in many electrical applications.

Semiconductors often show a *negative temperature* coefficient of resistance: that is, their resistance increases with decreasing temperature.

Kirchhoff's laws

Kirchhoff's laws greatly simplify the analysis of electrical circuits, and we shall have occasion to use the laws later. There are two laws, one applicable to current and the second applicable to voltage:

- the algebraic sum of the currents flowing into any junction is zero (see Fig. 1.2(a)); and

- the algebraic sum of the potential differences around any complete loop in a circuit is zero (see Fig. 1.2(b)).

The meaning of the words 'algebraic sum' means that account has to be taken of the signs of the quantities, so that current flowing out of a junction is counted as negative, and into the junction as positive. The current law is a consequence of the conservation of charge, which cannot be stored in a conductor, and the voltage law is essentially a statement of the conservation of energy, since the work done in moving a charge round a closed loop, where the starting and finishing potentials are the same, must be zero. It is most convenient to illustrate the use of these laws by tackling a real problem that

(a)

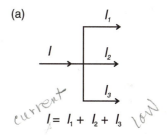

$$I = I_1 + I_2 + I_3$$

(b)

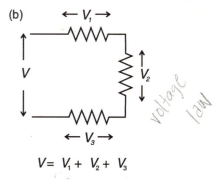

$$V = V_1 + V_2 + V_3$$

Fig. 1.2 Kirchoff's laws: (a) current law; (b) voltage law

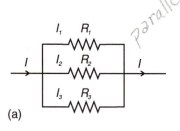

parallel

(a)

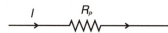

series

(b)

R_s

Fig. 1.3 Resistors in (a) parallel and (b) series

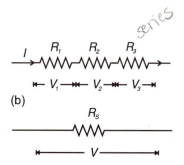

amplitude is proportional to the sine of the angle

Fig. 1.4. Variation of amplitude with time for a sine wave The angles marked emphasize the cyclic behaviour, and show that the sequence repeats itself after 360°

is of practical importance. Resistors may be connected in *parallel* or in *series*, as illustrated in Fig. 1.3(a) and (b) for a system of three resistors, R_1, R_2, and R_3. The problem is then to calculate the effective combined resistance, R_P or R_S, of the connected components. The application of the current law to the series situation, and the voltage law to the parallel one, shows rapidly that the results are.

$$R_P^{-1} = R_1^{-1} + R_2^{-1} + R_3^{-1} \ldots \tag{1.4a}$$
$$R_S = R_1 + R_2 + R_3 \ldots \tag{1.4b}$$

1.3 Direct and alternating currents

An important distinction is drawn between voltages and currents that remain of the same sign at all times and those whose polarity changes and alternates with time. It is usual to refer to the two types as *direct current (d.c.)* and *alternating current (a.c.)*, regardless of whether it is voltages or currents that are involved. The significance of the division into d.c. and a.c. is that many circuits respond in different ways to them. For the time being, however, we shall examine some of the essential features of a.c.

Sine waves

Figure 1.4 is a representation of an a.c. waveform that plays a peculiarly central role in all physics. It is the *sinusoidal* or *sine wave*. The figure may be thought of as graph of voltage across two terminals, or of current flowing in a circuit, plotted as a function of time. The ordinate (*y*-axis) is the time-dependent *amplitude* of the quantity being measured. The waveform is evidently recurrent or *cyclic*. Because of this cyclic behaviour, it is often convenient to regard the abscissa (*x*-axis) as showing how far round the cycle the waveform has progressed. Figure 1.4 illustrates the idea by showing that the waveform has repeated itself after 360 degrees, which is the angle of rotation in going round a full circle or cycle. This particular waveform is a sine wave because the amplitude is proportional to the sine of the angle. That is, $A = A_{max} \sin \theta$, where A is the amplitude at any time, A_{max} is the maximum amplitude at the crest of the wave, and θ is the angle progressed round the cycle. Remember that $\sin(180 + \theta) = -\sin \theta$, and you will see why the second half of the cycle is of opposite polarity to the first, and remember further that $\sin(360 + \theta) = \sin \theta$, and you will see that the behaviour is entirely consistent with the idea of repetition of the waveform after a cycle of 360 degrees. Although these angles have been expressed here in degrees, it is more convenient mathematically to employ units of radians. For those not familiar with these units, a brief explanation is given in Fig. 1.5. The feature to remember is that 360 degrees converts to 2π radians. One further property is of prime significance in dealing with a.c. phenomena, and that is the *frequency, f*. This frequency is simply the number of cycles that fit into each second, and is measured in *hertz*, abbreviated Hz. Put another way, the *period*

$$\frac{1}{f} = \frac{sec}{cycle}$$

$$f = \frac{cycles}{sec}$$

$$P = \frac{1}{f}$$

$$360° = 2\pi \text{ radians}$$

Change
degrees to
radians

$90° = \frac{\pi}{2}$
≈ 1.572

between successive cycles is $1/f$, and the total angle rotated in a time t is $2\pi ft$ radians (or $360ft$ degrees). The quantity $2\pi f$ appears time and again in a.c. formulas, which is one reason for introducing angular measurements in radians in the last paragraph.

The superposition of sine waves

We must now examine why the sine wave is so important in discussions of alternating waveforms. It is simply that all other waveforms can be produced by superposing an appropriate series of sine waves. This property was discovered by Jean Baptiste Fourier, and *Fourier synthesis*, *Fourier analysis*, and *Fourier transforms* are all physical or mathematical techniques or tools that exploit Fourier's discovery. Fourier synthesis is the building up of a more complex waveform through the superposition of sine waves. The total voltage or current is the algebraic sum of the individual sine-wave components, an idea probably familiar in the context of constructive and destructive *interference*. Fourier analysis is the reverse process of finding out what sine waves go to make up the more complex wave, while Fourier transforms are a mathematical tool used in performing this analysis.

Periodic waveforms

Let us start by considering waveforms that are repeated continuously in the same way that we have presumed the sine wave to be. Such repeated waveforms are called *periodic*. Two simple examples are the square wave (Fig. 1.6(a)) and the triangular wave (Fig. 1.7(a)). It turns out that the only sine waves needed to generate *any* such repeated waveform are those which are harmonics or multiples of some fundamental frequency f_0: that is, only frequencies f_0, $2f_0$, $3f_0$, $4f_0$, and so on, are needed. Indeed, for the square and triangular waves, the only frequencies required are the *odd* harmonics, f_0, $3f_0$, $5f_0$, ... Figure 1.6(b) shows how the addition of just the first two terms of this series makes the resultant waveform begin to look more like a square wave. Successive addition of the right amounts of the higher odd harmonics makes the resultant waveform approximate more and more closely to the square wave. An important phrase in the last sentence is 'of the right amounts'. The amplitudes have to be selected carefully to get the right result. For the square wave, the required amplitudes are in the ratio $1:1/3:1/5. . . . .$ Figure 1.6(c) illustrates the amplitudes needed for each harmonic in graphical form. The part played by the amplitude in determining the waveform produced by superposed waves is apparent when the triangular and square waves are compared. Although again only the odd harmonics are needed to synthesize the triangular wave, the amplitudes must this time be in the ratio $1:1/9:1/25. . . .$, as indicated in Fig. 1.7(b).

The two figures representing the intensities required at different frequencies, Figs 1.6(c) and 1.7(b), are the *spectra* of the two waveforms. The analogy with optical spectroscopy is complete. Light may consist of a

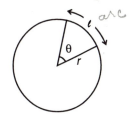

Fig. 1.5 Definition of an angle θ in radians: $\theta_{rad} = l/r$
For $360°$, $l = 2\pi r$ and $\theta = 2\pi$
For $90°$, $l = \pi r/2$ and $\theta = \pi/2$

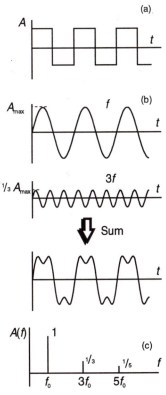

Fig. 1.6 Fourier synthesis of a square wave: (a) the square wave; (b) the first two components and their sum; (c) the amplitude spectrum

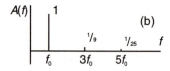

Fig. 1.7 Fourier synthesis of a triangular wave: (a) the waveform; (b) the amplitude spectrum

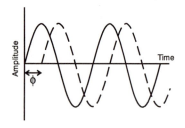

Fig. 1.8 Meaning of the phase angle ϕ for a single frequency sine wave

mixture of monochromatic wavelengths (single frequencies). Some device such as a prism or grating may be used to analyse these wavelengths or frequencies, in order to discover the intensity of each. The result is a spectrum. The opposite procedure, that of mixing light of different colours corresponds to the synthesis of a more complex waveform. In the nomenclature of Fourier transforms, which we shall discuss later, Figs 1.6(a) and 1.6(c), or Figs 1.7(a) and 1.7(b), are representations of conjugate *domains*, in this case the time domain and the frequency domain: that is, the waveform and its spectrum. The time-domain and frequency-domain figures contain the same information expressed in different ways. It is worth noting that a simple continuous sine wave corresponds to just one frequency and thus to a single line in the frequency domain.

Our two examples were chosen for their simplicity. In other cases, more complicated mixtures of amplitudes for different frequencies are required, but the principle remains the same. Sometimes the waveforms for the different frequencies must be displaced in origin from each other. Such a displacement is called a *phase shift*, and corresponds to a certain angle, ϕ, of offset for a given waveform (see Fig. 1.8). In terms of our equation for an individual sine wave, the equation $A = A_{max} \sin \theta$ becomes modified to $A = A_{max} \sin (\theta + \phi)$, and, in the synthesis of the complex waveform, each component may have its own individual value of ϕ.

Non-periodic waveforms

The statements made in the preceding paragraphs that all waveforms can be synthesized by the superposition of *harmonically related* sine waves, with correctly chosen amplitudes and phases has always been qualified by the requirement that the more complex waveform be repeated infinitely in time. The question must then arise about the feasibility of similarly representing a truncated waveform that is not repeated for ever, or even a pulse. The answer is that *any* waveform can be synthesized by the superposition of sine waves, but that the frequencies of these components will be a simple harmonic series only if the complex waveform is repeated indefinitely. In all other cases, the sine waves required consist of a continuous spread of frequencies rather than the individual 'monochromatic' values encountered so far. In the terms of spectroscopy, the discrete line spectra that correspond to the infinitely repeated waveforms have become replaced by a continuum. The reason for the difference between the sine-wave components of infinitely long wavetrains and those limited in time is not hard to see. A waveform synthesized from a finite number of discrete components is bound ultimately to repeat itself. The more components there are, and the closer they are together, the longer it takes for the repetition to occur. Only if there is an infinite number of frequencies will the pattern be confined to one period of time when all the sine waves interfere constructively, or are 'in phase'. At earlier and at later times, there is inevitably destructive interference between the infinite number of sine waves. Fig. 1.10 attempts to illustrate the way in which the superposition of a wide range of frequencies produces a waveform that dies away on either side of a transient region in the centre; this region becomes narrower as the range of frequencies employed to synthesize it becomes wider. What cannot be shown

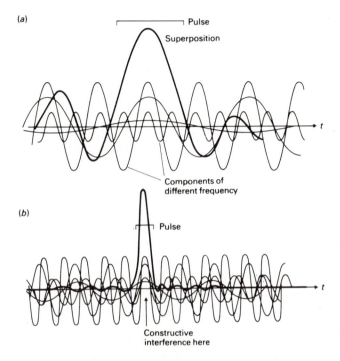

Fig. 1.9 Non-periodic waveforms: (a) a pulse; (b) a burst of oscillations

Fig. 1.10 A pulse has Fourier components of many frequencies needed for destructive interference at all times before and after the pulse. A narrower pulse (b) demands a wider range of frequencies than the broader pulse (a)

in such a diagram is that the frequencies employed must make up a continuum in both cases. It is, however, very easy to show the spectrum in the frequency domain corresponding to the two waveforms in the time domain, and the two pairs of representations are given in Fig. 1.11. A voltage pulse corresponds to yet another kind of spectrum in the frequency domain. As the pulse gets shorter, so the spread of frequencies gets larger, until in the limit of an infinitely narrow spike, all frequencies are represented at the same amplitude. The spectrum is then said to be 'white' by analogy with the visible spectrum of white light that contains all colours mixed together. At the other extreme, an unvarying d.c. voltage corresponds to an infinitely narrow spectrum containing only zero frequency. To conclude the sequence of time and frequency domain pictures, it is worthwhile including the pure continuous sine wave with which we started out.

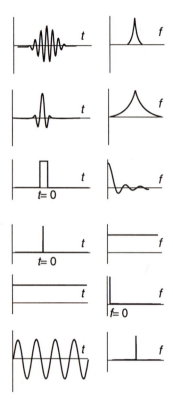

Fig. 1.11 Some waveforms (time domain) and their Fourier transforms (frequency domain)

The consequences of non-sine wave voltages

Many of the ideas presented in the previous sections have applications and consequences that will become apparent as the theme of this book develops. One aspect is in understanding how Fourier transform spectroscopy works. However, there are some other comments that it is convenient to make here. First, it is possible to interpret the response of many electrical and electronic circuit elements to complex applied voltage waveforms by considering the response to each individual sine wave component and summing the result. Some circuit elements are *non-linear* in their responses, in which case they produce an output which is not a pure sine wave and which thus contains harmonics. Secondly, it is evident that the electronic devices used in any instrumentation must be capable of responding to the highest frequencies present in a complex waveform if the output is not to be distorted from the input. A typical example that can readily be understood now is the frequency response demanded of amplifiers intended to increase the voltage of a pulse. Because, as we have just seen, a short pulse is equivalent to a very wide range of frequencies, the *bandwidth* of the amplifier must be sufficient to allow the pulse to be transmitted without information being lost. We shall return to this question of information and bandwidth shortly.

Since this is a book for chemists, it would be remiss of us not to look at another consequence of the frequency spread of short wavetrains. We know now that the shorter the train, the larger the spread. It is this behaviour that leads to the width of spectral lines being larger for transitions in which the excited state has shorter lifetimes. The exact relation between lifetime and linewidth can be determined from the quantitative forms of the information in time and frequency domains (see next section). Only a small step leads us to the *Heisenberg uncertainty principle*, which in its usual form links uncertainty in position and momentum of a (quantum) body or particle. Because particles also behave as waves, a localized particle corresponds to a wave that must be truncated because its position in space is known. The frequency is therefore spread out (uncertain), and since frequency and energy are related, the energy and ultimately the momentum are also uncertain.

Fourier transforms: a first glance

We have come, almost unwittingly, to a point at which the technique of Fourier transformation should be intelligible. Fourier transform methods play an increasingly important part in much of chemical instrumentation. They are used to improve enormously the sensitivity or resolution of several spectroscopic and related techniques. They are also used to tailor the behaviour of electronic devices, and they are used in the refinement and interpretation of the data obtained in the experiments. We shall return to a more detailed discussion of various kinds of Fourier spectroscopy, but it is appropriate here to look at the Fourier transform in the light of the material just presented. In outline, the Fourier transform is a mathematical technique that permits its user to convert information presented in the time domain so that it appears in the frequency domain, or *vice versa*. Other important pairs of domains are also connected by Fourier transformation, and they include distance and reciprocal length. The mathematical expressions involved are not unduly complicated, but do not really concern us yet. However, they can be applied analytically only to rather simple situations. The advent of high-speed computers has dramatically widened the scope of the methods, because numerical calculations of Fourier transforms can now be performed so rapidly as to be essentially in 'real time', and the operation is entirely routine. The really important point about the Fourier transform is that it consists of a summation (or integration) of sine functions corresponding in mathematics exactly to the addition of sine wave amplitudes, voltages, or currents that we have been considering in physics. As a result, the complex waveform in the time domain is identical to the Fourier transform of the frequency spectrum. One remarkable property of the Fourier transform is that, apart from some amplitude scaling factors and sign changes, if a function is transformed, and that transform is subjected to the transformation procedure a second time, then the original function is regenerated. Given that the waveform in the time domain is the first Fourier transform of the spectrum of the frequency domain, it follows that the second Fourier transform (properly speaking the *inverse* transform) of the waveform is the frequency spectrum itself. What all this means is that it is possible to shift simply and painlessly from one domain to the other.

As we have insisted repeatedly, the information contained in the time and frequency domains is identical, and that is made clearer now by seeing how the two are mathematically linked and how they can be interconverted. It is curious that most of us seem to think more readily of waveforms in the time domain than of the spectra in the frequency domain. Instruments exist that allow either domain to be examined. The *oscilloscope*, to be described in the next chapter, shows waveforms as a function of time, while the *spectrum analyser* allows the amplitudes of the frequency components to be displayed. The oscilloscope is, however, the more common device, and was certainly developed much earlier than the spectrum analyser; it is possibly these factors that colour our way of thinking.

Information and bandwidth

The reason for using any piece of instrumentation is to obtain information about an experiment. There is a precise quantitative definition of information (which is very closely related to the statistical definition of entropy), but the qualitative idea will suffice here. Information is conveyed in all sorts of different ways; examples include the speech or music in a radio programme, the picture on a television set, the data stream within a computer, the pattern on an oscilloscope, the trace on a chart recorder, and the movement of the needle on the scale of a meter. An underlying principle that governs the transfer of information is that it is only *changes* in the observed phenomenon that allow the information to be extracted. A truly continuous and unvarying response, such as a constant d.c. voltage or a sine or other periodic a.c.

The Fourier transform.

If a spectrum in the frequency domain is written $g(f)$, then a simplified Fourier transform $G(t)$ in the time domain can be written

$$G(t) = \int_{-\infty}^{\infty} g(f) \cos 2\pi f t \, df$$

and the inverse transform recovers $g(f)$

$$g(f) = \int_{-\infty}^{\infty} G(t) \cos 2\pi f t \, dt$$

(note that these simplified forms apply only to functions $g(f)$ and $G(t)$ that are symmetrical about $f = 0$ or $t = 0$; sin as well as cos terms arise in other cases).

waveform, cannot convey information. The reading on a meter provides information only if it is known that the reading is something different when the voltage is not present, and a tone heard on the radio conveys information only if a tone of different loudness (amplitude) or pitch (frequency) is known to be possible. In both these examples, and in every case besides, the changes required prevent the signal from being a pure d.c. or strictly periodic a.c. one. The more information that must be conveyed in a given time, the more rapidly must the amplitude or frequency be varied. On a chart recorder, the levels might change over a period of seconds or more, information conveyed by a voice corresponds to a few thousand variations in each second, while music has roughly ten times more information. A high-definition television picture needs several million changes of amplitude a second to be built up.

The treatment of non-periodic waveforms presented in previous sections should have made it clear that the alterations in the waveform needed to convey information will correspond in the frequency domain to a continuum of frequencies, the spread of which will increase as the rate of transfer of information increases. This spread is of the order of the rate of alteration of the waveform, and in our examples is tens of kHz (kilohertz = 10^3 Hz) for audible sounds and several MHz (megahertz = 10^6 Hz) for television pictures. There are consequences of the widened spectrum that have to be allowed for in scientific instrumentation and in all other systems that have to carry information. Electronic and other devices must be able to respond to the highest frequencies present, and they must be able to process the range of frequencies equally and without favouring some parts of the spectrum more than others. If such processing is not possible, then the signal will be distorted and information lost. The devices must therefore pass an adequate range of frequencies, in which case they are said to possess an adequate bandwidth. We met this term earlier in explaining the bandwidth demands of amplifiers for pulses, but the bandwidth requirement does not just apply to amplifiers and other electronic devices, but rather to the complete system. A meter or pen recorder can only register changes slower than the maximum rate at which the needle or pen can move, and a music system cannot convey to the listener frequencies beyond the capabilities of the loudspeaker.

It might be thought that suitable bandwidths could be attained, at least in principle, to handle any desired information rate, and that practical limitations might ultimately succumb to improvements in the devices. There is, however, a more basic limitation to the rate at which information can be handled. Noise is *always* present in any electrical system, because of the thermal motions of electrons and their statistical behaviour even if there are no external sources of noise. The reduction of noise is part of the challenge in inventing and improving chemical instrumentation, and Chapter 5 deals with the subject in detail. The signal has to compete with this noise, and the precision of measurements is degraded as a result. The extent of the degradation is determined by the relative contributions of the signal and the noise; this relative magnitude obtains quantitative expression in the *signal-to-noise ratio (S/N)*. Unavoidable noise in electrical systems is 'white'; that is, its amplitude is the same at all frequencies. The signal of interest is usually confined to a relatively narrow band about a fixed frequency, the bandwidth being determined by the information transfer rate as just discussed. An increase in the S/N ratio can thus be achieved by making the bandwidth of the signal processing devices as small as possible consistent with the desired information rate. Further increases in the S/N ratio can only be achieved if the information rate is reduced. There is thus a trade-off between the rate at which information can be transferred and the precision with which it can be obtained. Perfect precision is possible only if no information at all is transferred! The correlation between the uncertainty of the measurements (that is, the inverse of the S/N ratio) and the number of measurements that can be made looks very much like an uncertainty principle, and it arises for the same reasons as the quantum mechanical one. A fundamental limit is placed on the information transfer rate and the S/N ratio. Whether or not that limit affects the viability of an experiment depends, of course, on the signal level generated in the experiment and on the minimum precision acceptable in it as well as on the rate at which it is desired that the observations be made.

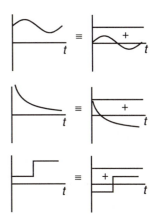

Fig. 1.12 Some complex waveforms can be regarded as the sum of a constant d.c. term and a time-dependent term

Combined d.c. and a.c.

Although we have not discussed the question so far, many waveforms encountered in scientific experiments have a constant term that has to be added to the time-dependent sine-wave terms in the Fourier synthesis. For our purposes, it is most convenient to think of such signals as being a combination of a mean d.c. level and an a.c. component (Fig. 1.12). In that case, the d.c. component is really constant, and all the information is conveyed by the a.c. signal. Even a slowly varying 'd.c.' signal, or one that is switched on or off, can be viewed in this way. If the information is carried by the a.c. alone, then it may be desirable to separate the a.c. and d.c. parts, and that can be achieved with the capacitors, inductors, and transformers that are introduced in the next section.

1.4 Capacitors, inductors, and transformers

In this section we discuss some very simple circuit elements whose response, in distinction to that of resistors, is dependent on the frequency of the applied voltage or current. Because of this frequency dependence, the devices to be considered behave in completely different ways according to whether the applied signal is d.c. or a.c.

Capacitors and capacitative reactance

In its simplest form, a capacitor merely consists of two plates separated by an insulating medium. Such a device cannot conduct d.c. currents because of the insulating gap. It can, however, store charge (and hence energy) because application of a potential difference across the plates allows electrons to be drawn from one plate and deposited on the other. The *capacitance C* is

Fig. 1.13 The symbol for a capacitor. Capacitance, C, is defined by the equation

$$C = q/V$$

where q is the charge stored in the capacitor and V the voltage across it The energy stored in the capacitor is $\frac{1}{2}CV^2$

For capacitors C_1, C_2, C_3, etc in series, the total capacitance C_S is given by the expression

$$\frac{1}{C_S} = \frac{1}{C_1} + \frac{1}{C_2} + \frac{1}{C_3}$$

while for the capacitors in parallel the expression is

$$C_P = C_1 + C_2 + C_3$$

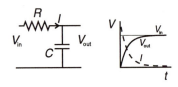

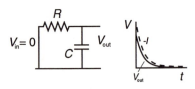

Fig. 1.14 The charging and discharging of a capacitor through a resistor

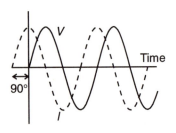

Fig. 1.15 The current and voltage waveforms for the circuit of Fig. 1.14 if the input voltage is an a.c. sine wave Note that the current 'leads' (is in advance of) the driving voltage by 90° ($\pi/2$ radians)

The capacitative reactance, X_C, is given by the expression

$$X_C = \frac{1}{2\pi f C}$$

defined as the ratio q/V where q is the stored charge and V is the voltage producing it. The basic unit used for C is the farad (F), although it is excessively large to be generally useful, and microfarads (μF), nanofarads (nF), and picofarads (pF) are more frequently encountered. Although some capacitors are constructed with air gaps, it is usual in practical capacitors to increase the amount of charge that can be stored by replacing the air by some other insulating material that has a high *relative permittivity* (sometimes called *dielectric constant*). Often, also, provision is made for the surface area to be increased, for example by making the plates of lengths of metal foil, and rolling them up with the separating dielectric material to make a compact capacitor. Capacitors joined together in parallel show a total capacitance equal to the sum of the individual values; in series, the capacitance is decreased.

The charge that may be stored in a capacitor has some direct applications for chemical experiments, because the charge can be built up relatively slowly, but discharged very rapidly. The photographic flash lamp is a familiar example of this use of stored charge, but, at higher energy levels, the discharge of capacitors is used in *flash photolysis* and in *pulsed laser* operation.

While a current cannot pass through a capacitor in the steady state, since charge is transferred when the potential of the plates is changing, there is certainly a transient current present whenever the potential changes. These ideas are illustrated in Fig. 1.14. There is thus a special situation when an alternating voltage is applied, because the capacitor is being continuously charged, discharged, and then charged to the opposite polarity. A current must then be flowing in the circuit at all times, because the steady-state situation is never reached. The capacitor appears to be acting as a conducting resistance in the circuit rather than as an insulator. The resistance is, however, rather peculiar because it is frequency dependent. For a given capacitance, the value is high for low frequencies where the rate of change of charge is small, while at higher frequencies this resistance becomes smaller. There is another important property of the relation between current and voltage that distinguishes the behaviour of the capacitor from that of an ordinary resistor. The current flowing is a maximum when the rate of change of voltage is highest, since that is when the charge also has to be shifted at the maximum rate. The rate of change of voltage is maximum at the crossing points of zero voltage for a sine wave (at 0, 180, 360 . . . degrees, or 0, π, 2π radians). At the crests and troughs of the applied voltage waveform, the voltage is momentarily constant, so that the current flowing is zero. In other words, the voltage applied to the circuit and the current flowing in it are shifted in time or phase with respect to each other, as indicated in Fig. 1.15, by one quarter of a cycle (90 degrees or $\pi/2$ radians). The current waveform precedes the voltage waveform, and the current is therefore said to *lead* the voltage by 90 degrees. For those who want to think a little more quantitatively about the behaviour, it is necessary only to remember that the rate of change of the voltage is its differential, that the differential of a sine function is a cosine, and that a cosine function leads its sine equivalent by 90 degrees. Because the behaviour of the capacitor towards a.c. is rather different from that of a true resistor, the special name *reactance* is given to the effective resistance. For a capacitor, it is usual to specify the quantity as *capacitative reactance* and to give it the symbol X_C; like resistance, it is measured in ohms.

Inductors and inductive reactance

A coil of wire constitutes an *inductor*, a device that is in many ways the counterpart of the capacitor. The coil is said to possess the property of *self-inductance*, which is given the symbol L, and is measured in henries. The behaviour of inductors is a consequence of the magnetic field that is produced when a current flows in a conductor. If a voltage is applied across a coil of wire, a magnetic field begins to grow. However, since this changing magnetic field is cutting a conductor, it induces a voltage in the conductor (a 'back e.m.f.') that opposes the voltage applied externally. As a result, the current flowing is initially small and builds up to its limiting value only as a steady-state is reached in the magnetic field. If the voltage is interrupted, the magnetic field collapses, again inducing a voltage in the coil that opposes the change of current flowing in the circuit. The magnitude of the effect will depend on the magnetic field produced and the induced voltage. For a particular geometry of coil, the inductance is proportional to the number of turns of wire. The inductance can be enhanced by inserting a suitable material with a high *permeability* in or around the coil; common materials used are soft iron and ferrite.

As with capacitors, the behaviour of inductors is most clearly demonstrated with a.c. voltages. If the conductor were perfect, at d.c. it would exhibit no resistance, and the current flowing in the circuit would be limited only by the resistance of the source. With a.c., however, the induced voltages reduce the current that flows, and the effective resistance of the circuit is increased over the true 'ohmic' resistance. The inductor is therefore exhibiting *inductive reactance*, to which the symbol X_L is given. This reactance *increases* with increasing frequency (compare with the behaviour of capacitative reactance, which falls with increasing frequency). The opposition to the change in applied voltage is maximum when the voltage is just going through zero and is changing magnitude most rapidly, and it is zero at the crests and troughs of the applied voltage waveform. As with capacitors, there is thus a 90 degree phase shift between voltage and current. In the case of inductors, the current does not reach its positive maximum until the applied voltage is passing through zero on its way from the positive to negative half cycle. The current thus *lags* the voltage by 90 degrees.

Circuits combining *L*, *C*, and *R*: impedances

We have seen that capacitors and inductors behave towards a.c. as though they possess a resistance that we have termed reactance. The reactance changes with frequency and imposes a 90 degree phase shift on the current with respect to the applied voltage. The question then arises of how to deal with

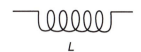

L

Fig. 1.16 The symbol for an inductor

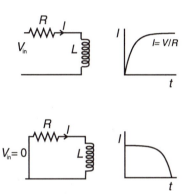

Fig. 1.17 Diagrams for the rise and fall of current in an inductor analogous to those shown for a capacitor in Fig. 1.14

In an a.c. circuit, the current 'lags' the voltage by 90°

The inductive reactance, X_L, is given by the expression
$$X_L = 2\pi f L$$

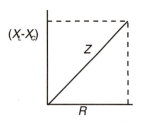

Fig. 1.18 A vector diagram showing how the complex impedance *Z* in a series circuit containing *C, L,* and *R* can be calculated from the capacitive and inductive reactances and the resistance in the circuit

circuits in which inductors, capacitors, and resistors are connected in series and parallel. If inductive (*L*) components, or capacitative (*C*) ones, are present alone, then the total reactance is obtained by using the formulas for resistance, eqns (1.4a) and (1.4b), substituting X_L or X_C for *R*. The combined circuit still behaves just as a combined *L* or *C*, and the phase relationships between current and voltage are those established already. If *L* and *C* are both present, then X_L and $(-X_C)$ can be used together, since the inductive current is 180 degrees out of phase with the capacitative current and subtracts from it. The situation becomes more complicated if resistance is present together with one of the reactive elements because of the phase shifts between the different current components. In fact, it is not difficult to show that the resultant effect of resistance and reactance is equivalent to an opposition to the flow of current which is the vector sum of all the reactive and resistive components. This net opposition is called the *impedance, Z,* of the circuit. To obtain the magnitude of the impedance it is necessary only to use Pythagoras's theorem (Fig. 1.18). For example, for *L, C,* and *R* all in series, the result is

$$Z = \{R^2 + (X_L - X_C)^2\}^{1/2}. \tag{1.5}$$

The phase can also be obtained from the vector diagram, but it is the magnitude of *Z* that is often most important. It is worth noting that the term impedance is often used when there are purely reactive or purely resistive elements present: the word is used to describe the opposition to the flow of current in a general way. Like resistance and reactance, it is, of course, measured in ohms.

Circuits combining *L, C,* and *R*: the frequency response

One of the most important uses of capacitors and inductors is in frequency-dependent circuits. At its most simple, this use can be looked on as a means of separating a.c. and d.c. signals. Consider a mixed a.c. and d.c. signal (Fig. 1.19(a)) applied to a resistor, but through a capacitor (Fig. 1.19(b)). The capacitor passes the a.c. component but not the d.c. one (Fig. 1.19(c)), so that the a.c. has been separated from the d.c. Made up the other way round, with the signal applied to a capacitor via a resistor (Fig. 1.19(d)), only the d.c. remains (Fig. 1.19(e)), because the capacitor acts to shunt away the a.c. component. These circuits constitute a *high-pass filter* and a *low-pass filter* and are very important building blocks in instrumentation. How well the a.c. component is retained in the high-pass filter or rejected in the low-pass one is a function of the component values and the frequency. The characteristic charging or discharging time in circuit 1.19(b), the so-called time constant, is equal to *RC*, and the frequency $1/(2\pi RC)$ marks a point at which the behaviour of both high- and low-pass filters alters markedly. Frequencies much higher than this point are passed without attenuation through the high-pass filter, but lower frequencies are attenuated by a roughly constant amount for each halving of frequency (Fig. 1.20(a)). The low-pass filter shows converse behaviour (Fig. 1.20(b)).

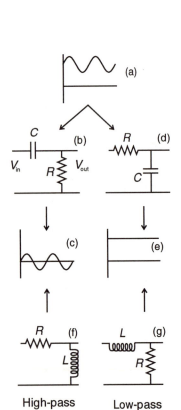

High-pass filters   Low-pass filters

Fig. 1.19 High-pass and low-pass filters

Inductors can also be used to make up high- and low pass filters, as shown in Fig. 1.19(f) and (g). Consideration of the frequency-dependent reactance of inductors will show that the parallel circuit (Fig. 1.19(f)) is this time the high-pass filter.

A combination of inductors and capacitors can likewise be used to generate desired frequency-dependent characteristics in a hybrid circuit. The most interesting behaviour of these circuits comes about because the current lags the voltage by 90 degrees in the inductor, but leads it by 90 degrees in the capacitor, as we noted earlier; eqn (1.5) shows that, without resistance ($R = 0$), the net impedance of an inductor and capacitor in series is $(X_L - X_C)$. But what happens if $X_L = X_C$? The reactance must go to zero and the inductor–capacitor pair present a short circuit to the driving voltage. The system is said to constitute a *resonant circuit*. The frequency at which resonance occurs is obtained by substituting the formulas for X_L and X_C. A parallel *LC* circuit is also resonant at the same frequency, but presents *infinite* impedance at resonance. Series and parallel resonant circuits can thus be used to construct filters that reject the resonant frequency (*notch filters*) or that pass the resonant frequency alone (*pass filters*), as illustrated in Fig. 1.21. Exactly these kinds of *tuned circuit* are used in a radio receiver to allow the signal of interest to pass and to reject all others. For purely reactive inductors and capacitors, the resonance would be infinitely sharp and the impedance at resonance truly zero for the series circuit or infinity for the parallel one. Because there are always resistive components in real components, this hypothetical situation does not occur in practice. A quantity Q, the *quality factor*, is a measure of how closely the practical circuit approaches the theoretical one. A similar Q-factor is encountered in many other situations where resonance is involved, ranging from violin construction to laser technology.

The properties of the filters just described can be improved in various ways that involve more complicated circuits, perhaps incorporating amplifiers. The rate of change of attenuation with frequency, and the sharpness of the 'knee', in the high- and low-pass filters can be modified for different purposes. Notch and band-pass tuned circuits can also be constructed from *RC* components alone. In all cases, however, the principles just set out lie behind the more sophisticated designs.

Transformers

The transformer is another very important and frequently encountered component. In essence, it consists of two coils of wire arranged so that a magnetic field produced by passing a current through one coil is also experienced by the other coil. If an alternating current is passed through the *primary* coil, the alternating magnetic field will induce a voltage in the *secondary* coil. With d.c. passing through one winding there is no induced current in the secondary because the magnetic field is constant. The transformer can thus be used to isolate completely an a.c. signal from a d.c.

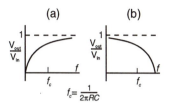

Fig. 1.20 Frequency responses of the high- and low- pass filters: (a) for the circuit of Fig. 1.19(b); (b) for the circuit of Fig. 1.19(d)

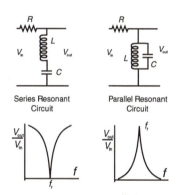

Fig. 1.21 Series and parallel *L-C* resonant circuits. The frequency of resonance, f_r is given by the expression

$$f_r = \frac{1}{2\pi(LC)^{1/2}}$$

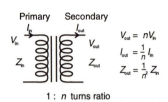

Primary   Secondary

$V_{out} = nV_{in}$

$I_{out} = \frac{1}{n} I_{in}$

$Z_{out} = \frac{1}{n^2} Z_{in}$

1 : *n* turns ratio

Fig. 1.22 A transformer, showing how the ratio of the number of turns of wire in the primary and secondary windings determines the relationships between input and output voltages, currents, and impedances

one. It has, however, other even more useful properties. If the ratio of turns of wire on the secondary to those on the primary winding is $n{:}1$, then $V_{out} = nV_{in}$, where V_{in} is the voltage applied to the input and V_{out} that produced at the output. Thus a transformer with a turns ratio greater than unity can be used to increase voltages (although the available current is reduced in proportion to the turns ratio). With a turns ratio of less than unity, a step-down transformer reduces the available voltage but the available current is increased. One further useful property of transformers is that, if a load of impedance Z_L is connected across the secondary, the primary behaves as though it had an impedance of Z_L/n^2. The device therefore transforms voltage, current, or impedance. Very great use is made of transformers at low frequencies for the distribution of mains power and in the fabrication of power supplies for experiments and for the instruments themselves. For these low-frequency purposes, transformers almost always have a high-permeability core coupling the two windings. But there are many other uses of transformers in interfacing experimental systems to the signal conditioning equipment. It is often the impedance-matching capability of the transformer that is of greatest value in these applications. For low-power, high-frequency purposes, the transformer may be air-cored.

1.5 Diodes, transistors, and integrated circuits

This section deals with a variety of *non-linear* electronic devices that are essential elements in the signal processing and conditioning offered by modern instrumentation. Their functions range from the simplest two-terminal component to devices that offer full-scale computing power. Electronics has its origins in investigating the behaviour of electrons in near-vacuum conditions, and *thermionic vacuum tubes*, or 'valves', were formerly used almost universally in electronic equipment. Semiconductor devices have, however, virtually replaced vacuum tubes except for a very few specialized applications, and we shall discuss only semiconductors here. They depend for their action on a semiconducting material, often silicon or germanium, *doped* with a small amount of impurity that provides excess electrons (n-type material) or a deficiency of them (p-type material). Our objective is to see what these devices do rather than to explain how they work, and the reader is referred to any one of a number of excellent texts that show why junctions of p- and n-doped semiconductors behave as they do.

Diodes

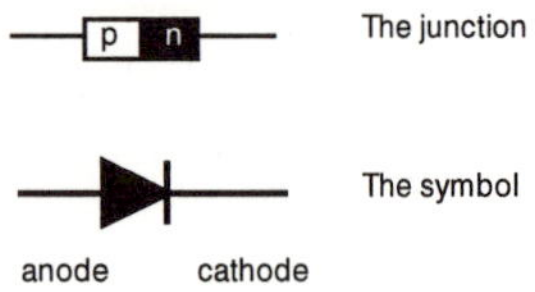

The junction

The symbol

anode   cathode

Fig. 1.23 The diode

The simplest of all the devices is the *diode*, and it consists of a single junction between p- and n- type material. One of the most important properties of the diode is that, expressed crudely, it allows electrons to pass from the n-side to the p-side, but not vice versa. Thus, if the n-side is made negative and the p-side is made positive, a current will flow, but if the polarity is reversed, the diode will behave almost as an insulator. If the diode is connected to a d.c.

voltage such that it is conducting, the negative terminal is the *cathode* and the positive terminal the *anode*. Of course, a real diode is neither a perfect insulator nor a perfect conductor, as the *I–V* graph shows (Fig. 1.24). For a silicon diode, conduction in the forward direction does not start until a threshold of about 0.6 volts has been passed, but beyond that it becomes a rather good conductor. The diode therefore behaves as a one-way element. One use is in converting a.c. into d.c. signals. If an a.c. signal is connected to a diode, only the positive half cycles will be conducted. A series of positive bursts of current or voltage appear at the output. A smooth d.c. level can then be obtained by connecting this output to a low-pass *RC* filter. This kind of application of the diode is called *rectification* and the diode is being used as a *rectifier*. Rectification circuits are used very widely in power supplies for instrumentation as well in converting a.c. signals that originate in the experiments, or that have been introduced in the instrument, into d.c. voltages for display and recording. Another use of diodes that makes use of their one-way conductivity is as high-speed *switches*, in which a *forward* or *reverse bias* is applied in order to make the diode allow or prevent the passage of other signals.

A rather different kind of application of diodes, but nevertheless an important one, is as a *voltage regulator*. The diagram showing the diode characteristics indicates that, if the diode is reverse biased, at some voltage it begins to conduct again. *Zener diodes* are manufactured so that this breakdown occurs very sharply and at a closely specified voltage. After breakdown, the current increases very rapidly for very small increases in voltage. If a voltage is applied via a resistor to a zener diode, then the voltage across the diode remains virtually constant regardless of the applied voltage so long as it exceeds the breakdown voltage. The zener diode can therefore provide a nearly constant voltage to an experiment or a piece of instrumentation regardless of variations in the supply voltage or in the load imposed (changes that are, perhaps, a consequence of altering current demands of the equipment), at least within certain limits. This property is clearly very useful, and is frequently used either directly or as part of a rather more sophisticated stabilizing circuit.

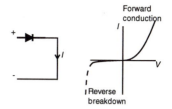

Fig. 1.24 Diode characteristics: relation between current and applied voltage in forward and reverse directions

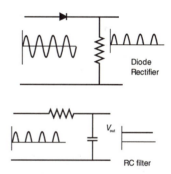

Fig. 1.25 The diode as a rectifier to convert a.c. into d.c. The lower diagram shows how the half-wave rectified output can be 'smoothed' by an *R-C* filter

Transistors

The next device to be considered is the three-terminal *transistor*. Two main types of transistor exist, the *junction transistor* and the *field effect transistor (FET)*. Diagrams for each are shown in Fig. 1.27; the three terminals are the *emitter*, *base*, and *collector* in the junction transistor and the *source*, *drain*, and *gate* in the FET. The way in which they are used is similar, and we shall confine our discussion for the time being to the junction transistor. If the semiconductor materials are arranged so that the emitter and collector regions are constructed of n-type material, and the base of p-type (see p. 16), the device is an *npn transistor*; the alternative construction is a *pnp transistor*.

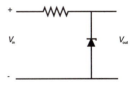

Fig. 1.26 The zener diode

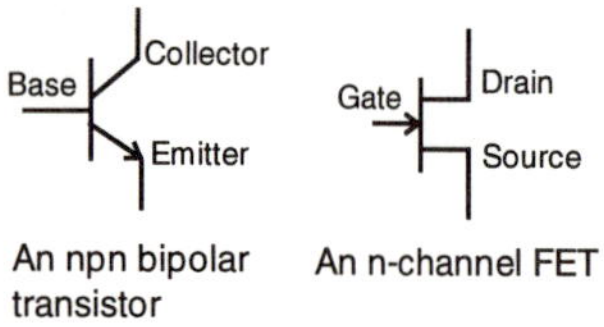

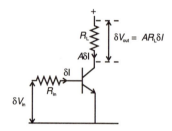

An npn bipolar   An n-channel FET
transistor

Fig. 1.27 The symbols for
(a) a bipolar transistor; and
(b) a field-effect transistor

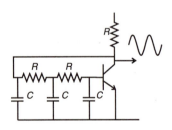

Fig. 1.28 A bipolar transistor
used as an amplifier. The
input current to the base
appears multiplied by the gain
factor *A* at the collector. The
circuit is that of a current
amplifier, but voltage
amplification can be achieved
by allowing the output collector
current to develop a voltage
across the load R_L

Fig. 1.29. An oscillator. The
R-C circuits shift the phase of
the output to be in-phase with
the input at the oscillation
frequency

The basic function of the transistor is to act as an *amplifier*. If the emitter
and collector are connected to a voltage supply, the transistor can pass a
current. To pass the emitter–collector current, it is necessary that a small
current is also injected into the base. However, the emitter–collector current
is strongly dependent on the base current, and changes by large amounts for
relatively small changes in base current. A change in base current δi appears
at the collector as a change $A.\delta i$, where A is the amplification or *gain* of the
transistor, and may be much greater than unity (values of several hundred are
not uncommon). We thus have the basis of a *current amplifier* which permits
a small current from some experimental system to be converted to a larger
current that can do something 'useful'. In one familiar example from the
domestic field, the small current might come from the head of a tape recorder
that converts the magnetic information on the tape to an a.c. signal, and the
large amplified current might drive the coils of a loudspeaker so that we can
hear the music. The transistor can also be used as a *voltage amplifier*, as
shown in Fig. 1.28, because applying a voltage change to an input resistor
connected in series with the base produces a change in base current. The
collector current can, in turn, be passed through a load resistor, and the
amplified changes in the resistor will develop amplified voltages across it in
accordance with Ohm's law. The FET works in a very similar way; one key
difference is that it is the input voltage, rather than an input current, that
directly leads to the change in output current. As a result, the input
impedance of an FET is very high (the impedance at the base of a junction
transistor is low), and this may be a very desirable quality if the amplifier is
not to 'load' the signal source and thus alter its characteristics.

To obtain higher gain, many transistor amplifiers can be connected together
in suitable ways. Many other important circuit functions can also be obtained.
Sometimes it is necessary to generate sine, square, or other shaped waves at
a specified frequency in order to drive some experiment. One or more
transistors can be configured to act as an *oscillator* (Fig. 1.29). Some of the
amplified output is fed back to the input in the correct phase so that it is also
amplified. If the connections from output back to input are made *via*
frequency-dependent circuitry (see Section 1.4), then the complete device may
produce sustained oscillations at the desired frequency. The transistor may
also be used as a *switch*, because with suitable base current or gate voltage,
the transistor can be made to pass negligible current, while with other currents
or voltages, the transistor can be made to appear as a fairly low resistance.
Under these conditions, the transistor is said to be *saturated*, and the output
current is no longer much dependent on the exact input conditions. The
transistor is either 'off' or 'on'. The currents it passes can be much larger
than those that control its switching function. Applications of this behaviour
are very numerous, and include switching on lamps, heaters, or motors.
Perhaps more important than any of these uses, however, is the possibility of
using the switching properties in *digital logic circuits*. Such circuits are at the
heart of computers, and will be described in the next section.

Integrated circuits

Nothing has been said about the detailed design of circuits incorporating transistors. One reason for this omission is the present-day widespread availability of *integrated circuits* that can perform almost any desired function with the addition of a few external components. Within the integrated circuits are anything between two and tens of thousands of transistors, together with the associated diodes, resistors and capacitors, often all laid out on the same small silicon wafer or chip. The internal design and component values have been optimized, so that the user does not have to worry about these aspects. Mass-produced integrated circuits are exceptionally cheap, and, except for special purposes, it is rare for the experimental chemist or physicist to have to use discrete transistors at all. The design and construction of instrumentation has thus altered from one in which the detailed electronics have to be considered, and is now more akin to fixing together a series of building blocks to perform the required operations.

There are two families of integrated circuits: *analogue* and *digital*. The analogue circuits are based on amplifiers. A continuum of input and output voltages or currents is possible (although the relationship of output to input is not necessarily linear). Amplification itself is obviously a central application of these integrated circuits, but many are designed with more esoteric specific uses in mind, including fabricating oscillators, comparators, multipliers, and so on. We shall examine a special use of amplifiers as *operational amplifiers* in Chapter 4, and defer further discussion of linear and analogue circuits until then. The digital circuits are based on the switching capability of the transistor, and respond only to whether a signal is present beyond a threshold level. The input levels in many of these devices are thought of as logical '0' or '1', and the output is also usually just '0' or '1' as well. A typical simple integrated logic element is the *NAND gate* (the word 'NAND' stands for 'Not AND'). The *truth table* for a NAND gate is shown in Fig. 1.31. Other important logic gates (many of which can be built up just from NAND gates) include those with *AND* and *OR* functions. These simple elements can then be used to construct more complex logical functions and are at the heart of counters, dividers, memories, and much more besides. Of course, more complicated integrated circuits are fabricated that combine many of these functions on single chips, and they reach a very high level of sophistication in the *central processing unit (CPU)* of even the most humble computer. The computer itself is beginning to replace the individual integrated circuits in many instrumentation applications, because it can be *programmed* to perform a multitude of functions that would previously have required dedicated integrated circuits or large arrays of logic chips for which the design would have demanded great thought and effort. At least at the level of a CPU and associated memory and peripheral integrated circuits, it is not unreasonable to provide a small programmed computer to replace many of the specialized devices. Of course, it is a far cry from the days when a computer itself would require rooms full of vacuum tubes, but then that is the kind of progress that has made exciting developments in chemical instrumentation possible.

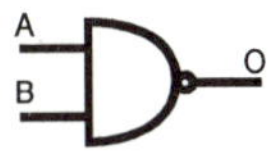

Fig. 1.30 Symbol for a NAND gate

A	B	O
0	0	1
0	1	1
1	0	1
1	1	0

Fig. 1.31 Truth table for the NAND gate. If both A and B inputs are at logical '1' levels, the output is '0'. For all other input combinations, the table shows that the output is '1' That is, the output is affected only if A *and* B are '1'; because this is a NAND gate, the output is *negated* (ie, '0' rather than '1')

2   Simple measuring instruments

2.1 Input and output transducers

The objective of the instrumentation described in this book is to convert some phenomenon of chemical interest to an electrical signal, then to process and manipulate that signal, and finally to turn the processed result into something useable by the experimenter. In this chapter and the next, we examine the output and input interfaces with the experiment and the experimenter. The devices that provide the interface are *transducers* because they convert a non-electrical phenomenon to or from an electrical one. In modern instrumentation, of course, the final link may be a computer and its associated memory and storage, so that the interface with the investigator may be the screen display, the printed results, or similar output.

We start by discussing the final link, the output transducer, since these devices are often in reality rather familiar electrical measuring instruments, such as meters, pen recorders, and oscilloscopes. Different senses of the observer may be called into play. The three devices just mentioned depend primarily on vision. Hearing may also be used, as, for example, in the detection of an audiofrequency a.c. signal. Before getting involved with the transducers themselves, it is useful to consider the types of electrical signal that they are called on to convert for interpretation by the experimenter.

Analogue signals

Electrical signals may often be in the form of a voltage, current, or, more rarely, charge or power generated by the chemical system under investigation. These signals may be converted from one form to another (e.g. from current to voltage) in the electrical part of the instrumentation, but the measured quantity remains the amplitude of one of the variables, and its variation with time or experimental conditions as explained on pp. 9–11. All values of the amplitude are essentially possible since the unit of 'quantization' is the tiny charge on a single electron, and voltage, current, charge, and power are all *continuous* quantities. Determinations of them are *analogue* measurements.

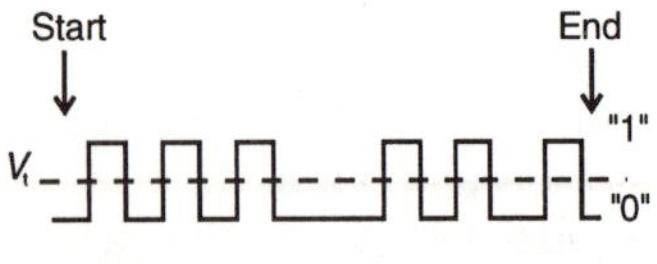

Digital signals

Information can be conveyed electrically in a quite different manner. The presence or absence of a voltage may define a logical 0 or 1 (see p. 19), and the data of interest are encoded in the pattern of 0s and 1s. For example, Fig. 2.1 shows a train of pulses that might be generated by a detector of radioactive particles. The quantity of interest is the total *number* of pulses detected between the start and the end of the measurement. Although some voltage must define the threshold between the two logic levels, shown in the

Fig. 2.1 A train of pulses that can be counted digitally. Six pulses have arrived between the start and the end of the measurement in this example

diagram as the dotted line labelled V_t, the magnitude of the voltages making up the pulse train are unimportant so long as the 0s are all below V_t and the 1s all above it. The signal is *digital* because it conveys a number; the minimum 'quantized' difference between successive numbers is evidently one unit (although it might be larger). Our example is of a series of sequential discrete events that are counted. Such a signal is referred to as *serial*. One single binary digit is encoded by the pulses. In computer and logic jargon, this single digit is called a *bit*. *Parallel* encoded digital signals are also frequently used. Instead of the data being sent bit-by-bit (a very appropriate expression in plain English, since it is also precise in logic terms), more data can be sent simultaneously over several wires. Consider the four-wire digital device of Fig. 2.2. Each of the wires can carry a 0 or a 1. If the wires are connected to indicators that read 1, 2, 4, 8 when activated, then the device can show any number between 0 and 15 (1 + 2 + 4 + 8). As set up in the figure, the number indicated is 11 (1 + 2 + 0 + 8). The minimum number change is still unity, and the *resolution* is one part in 16. If there had been eight parallel wires, the eight bits of information sent simultaneously would have been said to constitute one *byte*. (The eating analogy is not discarded for the four-bit case, which is a *nibble*, or half a byte). There is nothing to prevent a series of pulses being sent over several parallel lines more or less simultaneously, and that is what happens in computers and many other digital and logical devices. The specification of a computer as containing 'eight-bit' or 'sixteen-bit' parallel processing gives some indication of its capabilities, although speed of the operations is a key element as well.

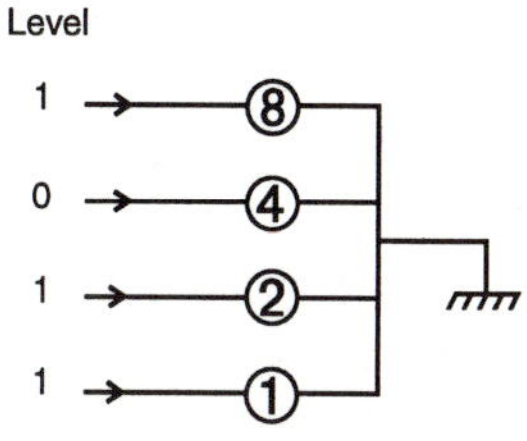

Fig. 2.2 Display of a number (11 in this case) from parallel encoded digital information

Time measurements

Yet another class of measurement uses information from the timing of signals. The timing operations are generally carried out on voltages present as logical 0s and 1s, although these signals might have started out as analogue voltages. The simplest measurements are those in which the time between two successive events is determined. This time might be the period between two transitions between logical 0 and 1, the *period* (t_p in Fig. 2.3), or the *pulse width* itself (t_w in the figure). Typical real-life examples of period and width are the time between successive trains passing a road crossing and the time taken for the individual train to go over the crossing.

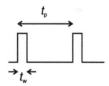

Fig. 2.3 The width of a pulse (t_w) and the period between pulses (t_p)

A slightly more involved use of time measurements is in determining the number of events that occur in unit time. If, in the example of Fig. 2.1, the 'start' and 'end' were separated by unit time (one second, one minute, or whatever other unit is chosen), then the measurement would have been of number of events per unit time, or the *rate*. The *frequency* of an a.c. signal is a very similar quantity. Here, it is necessary to know the number of cycles per unit time (usually one second), and thus the number of times the pure a.c. component (see Fig. 1.12, p. 11) crosses zero in the same direction. Figure 2.4(a) marks the crossings as the signal goes from negative to positive.

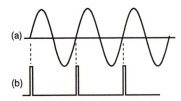

Fig. 2.4 Conversion of (a) an analogue a.c. signal to (b) digital logic pulses

Modern frequency measurements often start by converting an analogue a.c. voltage to a series of digital logic pulses at the crossing point (Fig. 2.4(b)), and then counting the number of pulses in a fixed time period. Finally, the number is divided by the time to give the frequency.

Consideration of the techniques suggested for measuring rates or frequencies will indicate that the accuracy of the determination increases with the length of the sampling period. In a very short period there might be no pulses to count, but that does not mean that the rate or frequency is zero. Conversely, one pulse might just occur in that short sampling time, and would translate to a high frequency, even though the true frequency might be quite low. In Fig. 2.1, the 'start' and 'end' may or may not just include the first and the last pulses. There is apparently no substitute for making measurements over a sufficiently long time that these sampling errors are reduced to an acceptable extent. An intimate connection exists between this problem and the impossibility of knowing the frequency of a truncated waveform that is not repeated for ever (see Chapter 1, pp. 6 *et seq.*)

Interconversion of analogue, digital, and time signals

Much modern instrumentation takes advantage of the increasing ease with which the three types of signal can be interconverted. Complex integrated circuits, in particular, have removed many of the difficulties associated with the conversions. It is thus often convenient to obtain the initial information from the experiment in one form (frequently analogue), manipulate it in the instrument perhaps in several forms, and final display it another (for example digitally). *Analogue-to-digital converters* (*ADCs*) and *digital-to-analogue converters* (*DACs*) are commonplace. An example of time-to-analogue conversion is found in the *time-to-amplitude converter* (*TAC*), which we shall encounter again in Chapter 7, and the *voltage-controlled oscillator* (*VCO*) has an output frequency that depends on input voltage, and furnishes an example of a device that converts analogue-to-time signals.

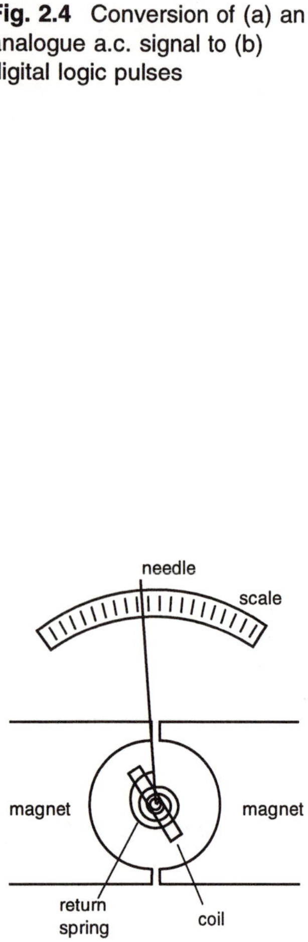

2.2 Analogue meters

The most familiar of all output transducers must be the simple meter. There are various constructions, but the only one with which we shall be concerned is the moving-coil meter. In essence, the meter consists of a coil of wire suspended between the poles of a magnet, and held against the action of a spring (see Fig. 2.5). When a current passes through the coil, the induced magnetic field opposing the field of the fixed magnet generates a repulsive force that is balanced against the tension in the spring. The extent of (angular) movement of the coil is thus dependent on the magnitude of current flowing in the coil. A needle is attached to the coil, and it moves against a scale, which the observer reads. By great good fortune, the displacement on the scale of a properly made moving coil meter is a linear function of the current flowing. The device is strictly analogue in behaviour, although the

Fig. 2.5 The elements of a moving-coil meter movement

resolution may be limited by the ability of the operator to read the position of the needle against the scale. To some extent, the result is subjective, since the skill of the experimenter enters into making the reading. The electrical quantity measured is the current flow, and the moving-coil meter is sometimes called an *ammeter* (or *milliammeter* or *microammeter*, depending on the current to be measured). The sensitivity of the meter is determined both by the field of the fixed magnet, and, more especially, by the number of turns of wire in the coil. Meters with large numbers of turns of fine wire may give a *full-scale deflection* for 100 microamps or less. Less delicate devices with fewer turns of thick wire can be used to measure currents running into hundreds of amps. It is, however, usual in instrumentation applications to measure currents in excess of a few hundred milliamps with a more sensitive meter equipped with a *shunt* (Fig. 2.6). Simple application of Ohm's and Kirchhoff's laws (cf. Chapter 1) allows the required parallel shunt resistance, R_p, to be calculated for any desired sensitivity, so long as the resistance of the meter, R_m, is also known.

Although the moving coil meter responds to currents flowing through the coil, it can be adapted to the measurement of voltages as well. For this purpose, a series resistance is incorporated in the circuit (Fig. 2.7). Once again, the simple electrical laws allow the value of the series resistance, R_s, to be calculated. If $R_m \ll R_s$, the current flowing is nearly equal to the applied voltage divided by R_s. Thus, for a meter that gives full-scale deflection for a current of 100 µA, a series resistor of $10^6 \, \Omega$ (1 MΩ) would provide a voltage sensitivity of roughly 100 V. For a sensitivity of 1 V, the value of R_s would have to be about $10^4 \, \Omega$ (10 kΩ). A real 100 µA meter might have a coil resistance of $10^3 \, \Omega$ (1 kΩ), so that the correct series resistance for an accurate reading of 100 V full scale would be $(10^6 - 10^3) \, \Omega = 999$ kΩ, and for the 1 V meter the corresponding value of R_s is 9 kΩ. Whatever the values of the series resistance needed, it is evident that the meter needs to draw some current to make it work, and the total resistance of the circuit, $R_s + R_m$, is imposed as a load across the voltage source being measured. For low impedance sources, this may not matter very much, but for sources where the impedance is comparable with or larger than the meter-circuit resistance, very considerable measuring errors may be introduced by the loading. Even if the meter resistance is ten times larger than the source resistance, there is an error of approximately ten per cent in the voltage measurement. The distinction between *electromotive force* (*e.m.f.*) and *potential difference* (*p.d.*) was glossed over in the first Chapter. The essential distinction is that e.m.f. refers to the load-free voltage produced by a source, while p.d. applies to the voltage under load, whatever that load may be. It follows that a moving-coil voltmeter is not capable of measuring true e.m.f.s, and that is one reason why it is not suitable for determining electrode potentials (pp. 48 *et seq.*) directly. Different instrumentation must be used that does not draw a current. Alternative methods must be used that either intrinsically draw no current (e.g. the potentiometer, p. 31) or that *buffer* the meter from the voltage source (see Section 2.3 below and pp. 56–57).

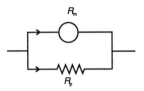

Fig. 2.6 Use of a shunt to increase a meter range to measure larger currents

The current flowing through the complete parallel circuit is a factor of $(R_p + R_m)/R_p$ times larger than that flowing through the meter itself, so that the meter range is increased by this factor

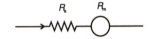

Fig. 2.7 Measurement of voltage with the moving-coil meter. A series multiplier resistor allows a current of $V/(R_s + R_m)$ to flow through the meter for an applied potential difference of V

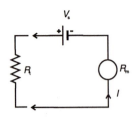

Fig. 2.8 Use of the moving-coil meter to measure an unknown resistance, R_t. A voltage source, of magnitude V_s, causes a current $I_m = V_s/(R_t + R_m)$ to flow in the meter. The value of R_t is thus given by the equation

$$R_t = \frac{V_s}{I} - R_m$$

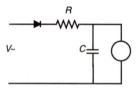

Fig. 2.9 Measurement of a.c. voltages with the moving-coil meter. A diode provides half-wave rectification, and a simple *RC* circuit smooths the output

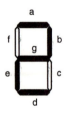

Fig. 2.10 A seven-segment display energized to show the figure 2

Resistance measurements are also possible with the moving-coil meter. What is required is a source of known voltage connected in the meter circuit as shown in Fig. 2.8. The current is measured, and is related to the resistance of the device under test in a simple (but non-linear) manner.

The moving-coil meter is polarity sensitive, because the magnetic field produced in the coil depends on the direction of current flow. The suspension of the coil can be arranged so that the needle points to a central position when there is no current flowing, and such a meter is capable of measuring currents (or voltages) of either negative or positive polarity. However, the moving-coil meter in unmodified form is not suitable for measuring a.c. currents or voltages, because the coil and needle movement cannot keep up with fluctuations on time scales shorter than tenths of a second to seconds. If the meter is to be used for a.c., some rectification circuit must be used to convert the a.c. to d.c. (cf. p. 17). Figure 2.9 shows a simple way of achieving the objective. Some care has to be taken in interpreting the voltages measured, partly because of the forward voltage drop of the diode in this circuit, but also because, depending on the *RC* time constants in the circuit, the meter may read the peak voltage of the a.c. waveform, or some averaged value. The average most often used in calculations is the *root-mean-square* (*rms*) value.

A very common electrical test instrument combines many of the functions just described in a *multimeter*. Switches or plugs enable the user to select measurement of d.c. or a.c. voltages or currents at several different sensitivities, and resistance ranges are often included as well.

2.3. Digital meters

Modern instrumentation relies heavily on digital displays, which have superseded analogue meters for many purposes. Some of the most convenient of the displays show decimal digits (0 to 9) directly, and can be placed alongside each other to allow any number of digits to be presented. These displays rely either on the light emitted by certain types of semiconductor diode (*light emitting diode*, or *LED*) when a current is passed through it, or on the rotation of the plane of polarization of light by certain liquid crystals when exposed to an electric field, which forms the basis of the *liquid crystal display*, or *LCD*. The numbers that are seen are made up as a matrix of dots or as a group of segments that are illuminated (LED) or that show up as dark against a lighter background (LCD). Figure 2.10 illustrates one common arrangement of seven segments, here showing the number two. The selection of which segments to activate is performed by an integrated circuit decoder, the principle of which is outlined in Fig. 2.11 for a four-bit parallel input where only the 'two' line is at logical 1 level, and the outputs are at logical 1 for the appropriate segments of the display (numbers beyond nine would have to be dealt with specially in this simplified illustration). Sometimes the display and decoder circuits are put together in one integrated circuit package.

So far, we have considered only the display aspects of these digital circuits. The display has to be preceded by processing of the input signal. The *digital voltmeter* (*DVM*) uses suitable amplifiers and an ADC (see p. 22) to convert the analogue input voltage to the required parallel digital logic levels. Circuits in common use at present often have very high input impedances (typically 100 MΩ for a 200 mV sensitivity). Currents drawn are thus of the order of nanoamps, and the system under test is virtually unloaded by the DVM, in contrast to the situation that arises with a moving-coil voltmeter (p. 23), and something approaching true e.m.f.s may be measured. One important point to note is that the resolution of the DVM is limited by the 'quantization' of the least significant digit; on the other hand, the reading of the numbers is objective and does not depend on any particular operator skill. With the addition of suitable resistor networks, currents can also be measured digitally, and, with a voltage source, resistance determinations can be made. Digital multimeters are now commonly found as laboratory testgear.

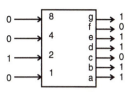

Fig. 2.11 Decoding of a parallel binary logic input to drive the appropriate segments of the display

2.4 Counters and timers

Digital displays have other valuable uses in providing an interface between instrumentation and the human user in *counters* and *timers*. The building block on which these instruments are based is usually a *flip-flop* (also sometimes known as an *astable multivibrator*) that changes its logic state from a 0 to a 1, or from a 1 to a 0, for every input pulse that it receives. A counter accepts a sequence of logic pulses to change the logic state of a flip-flop, and each time there is a transition from 1 to 0, a 'carry' pulse is sent to a similar binary device that counts the next significant (binary) digit, as illustrated in Fig. 2.12. A stack of such binary counters produces as many bits as required of an output that corresponds to the total number of input pulses. The binary output is then taken to the decoder (Fig. 2.11) and display (Fig. 2.10) to produce a readable value of the count. Maximum count rates are typically 100MHz (ie one hundred million counts per second). Many counters are equipped with 'start' and 'stop' controls, usually operated by electrical impulses supplied by the user.

Timers work in a similar way, except that the pulses are derived from a high-accuracy quartz-crystal controlled oscillator of known frequency. The 'start' and 'stop' signals now allow the counting of these internal pulses. Simple division of the total number of counts by the oscillator frequency gives the elapsed time, which can be displayed as described in the last paragraph (note that dividing by 2, 4, 8, etc merely requires shifting the connections to the display one binary digit higher up the chain of binary outputs from the counter). A *clock* made up of an electronic timer of this kind can also be used to generate the 'start' and 'stop' pulses of a second counter, which then forms the basis of a *digital ratemeter* or *frequency meter*.

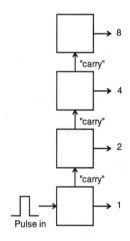

Fig. 2.12 A four-bit binary counter

2.5 The oscilloscope

The *cathode-ray oscilloscope* (*CRO*), a term usually shortened to just 'oscilloscope', is one of the most valuable of all electronic devices for use in scientific instrumentation and in testgear. Its value derives largely from its ability to display a.c. waveforms, including pulses and trains of pulses, in a way that looks like an amplitude–time graph, and that is thus capable of instant interpretation by the user. There are other uses besides, as will appear later in this section, and the *cathode-ray tube* (*CRT*) which is at the heart of the oscilloscope is also the key component of the *visual display units* (*VDUs*) associated with most computer terminals.

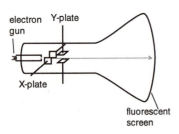

Fig. 2.13 The main elements of the cathode ray tube

Figure 2.13 shows the essential features of one type of CRT. Contained in an evacuated enclosure are an *electron gun*, two sets of electrostatic *deflection plates*, and a fluorescent target *screen*. A cathode heated by a filament produces electrons that are accelerated by a large potential difference and focused by further appropriate fields. Without any field applied to the deflection plates, this arrangement would produce just a single sharply defined spot where the beam of electrons hit the fluorescent screen. With a field across a pair of plates, the electrons are attracted towards the positive plate (and repelled from the negative one), so that the beam is deflected from its straight-ahead direction. One set of plates is arranged to give horizontal deflection (usually called the X-plates, in accord with the convention for labelling graph axes), while the other pair of plates, the Y-plates, is arranged perpendicular to the X-plates, and gives vertical deflection. Each pair of plates is normally connected to an amplifier so that small voltages applied to the inputs of the amplifiers move the spot substantial distances across the screen, the distance being proportional to the applied voltage.

Linear sweep display

In the most usual way that the oscilloscope is used, the X-plates are connected to a *time-base* that generates an a.c. voltage with a saw-tooth waveform like that shown in Fig. 2.14. The spot is driven by this voltage from left to right across the screen as the voltage increases up the ramp for the period τ, and then it immediately returns to its starting position at the left-hand side of the screen during the *fly-back* period, and the cycle starts over again. With a sufficiently slow time-base (long τ, the spot can be seen to move across the screen and fly back again repeatedly, but at time-base frequencies above about 10 Hz, the motion blurs out, and the eye perceives a straight horizontal line on the screen. Application of a d.c. voltage to the Y-plate amplifier will now shift the horizontal line up or down according to the polarity of the field. If, on the other hand, an a.c. signal is applied to the Y-plates, then the spot is moved up and down as it traverses the screen, and a pattern is seen that corresponds to the waveform of the a.c. applied. A sinusoidal a.c. signal produces a sine-wave pattern on the screen, and so on. The oscilloscope thus provides a way of 'seeing' the waveform of the a.c. The frequency of the time-base must, of course, be adjusted so that a suitable part of the waveform

Fig. 2.14 Waveform of a saw-tooth ramp voltage used to drive the X-plates of an oscilloscope

of interest, often many cycles, fits onto the screen. Laboratory oscilloscopes in common use have frequency responses covering a range from d.c. to 100 MHz or more, with the time-base frequency being selected by a range-switch and fine control. Not only can the presence of a.c. signals be detected and their waveforms examined, but the frequency can also be estimated if the time-base is calibrated. In a development of the simple oscilloscope, an additional electron gun can be added (or simulated electronically) to give a *dual-beam oscilloscope*, and even more beams are possible. The dual-beam oscilloscope allows two waveforms to be examined simultaneously, and the frequency and phase relationships between them can be investigated. The X–Y display to be discussed in the next section is, in fact, more useful for accurate measurements of frequency and phase of sine and related waveform signals, but the dual-trace oscilloscope is valuable for examining the amplitude and shape relationships between different signals, and is particularly useful in following the timing of pulse signals in various parts of a circuit.

One essential feature of an oscilloscope used in the linear-sweep mode is a means of synchronizing the start of each sweep with the waveform under examination. The details of how the *synchronization* or *triggering* are achieved need not concern us, but without the synchronization, each sweep would produce a pattern on the screen shifted at random with respect to its predecessor, and the eye would see only a muddle of lines. An important operator skill in adjusting an oscilloscope lies in being able to set up the triggering correctly.

Although we have described the linear-sweep display in terms of a repeated saw-tooth X-plate field, the oscilloscope can also be employed as a single-shot instrument with linear sweep. For example, the oscilloscope might be used to follow the intensity–time dependence of a single laser pulse, a suitable photodetector (see Section 3.6) being used to convert the light to an electrical signal. In this application, the start of a single sweep up the X-voltage ramp is triggered by some external event (such as an electrical pulse that also triggers the laser), and the electron beam traverses the fluorescent screen just once. It is evident here that the triggering is critical to the success of the experiment, since it is no good the spot having finished its travel, or not even having started it, when the laser fires. To make an accurate measurement of the time evolution of the light pulse from the moment of triggering the laser, it is essential that the relation between the start of the oscilloscope sweep and the initiation of the laser pulse be known exactly.

The use of the oscilloscope in the single-shot mode illustrates a difficulty that is associated with the recording of very fast transient signals that are not repeated. The event is over much more quickly than the eye can record it. One fairly simple way out of the difficulty is to make the fluorescent screen have a relatively long *persistence*, so that the glow remains for some time after the electron beam has passed by. Another is to photograph the screen, with the camera shutter being opened just before or as the triggering pulse is received. A whole variety of more sophisticated *storage oscilloscopes* has been devised that memorize, by one means or another, the transient waveform, and keep it available for display. Many of the storage oscilloscopes are, in reality, computers that store a digitized version of the signal in memories that are addressed sequentially by a time-base. The memory contents are then read out repeatedly and converted to an analogue voltage to give an apparently continuous display on the oscilloscope. We are entering the realms of using computers in data acquisition, which is touched on in Chapter 6; further discussion can be found on pp. 76–77.

X–Y display

An alternative valuable way of using the oscilloscope is to drive both the Y and the X plates with voltages derived from the experimental system, rather than energizing the X plates from an internal source. In this case, the oscilloscope gives a plot of the Y-voltage against the X-voltage rather than of the Y-voltage against time. However, if both the applied voltages are repetitive a.c. ones, then the oscilloscope spot repeatedly traces out the way in which one voltage varies with time as a function of the way the other one varies with time. If the frequencies of the two a.c. signals bear a fairly simple relationship to each other, then the pattern appears stationary. Such a pattern is called a *Lissajous figure*. Let us first consider the case where two sine wave signals of the same frequency are used. The Lissajous figure then reveals the phase *relationship* between the signals (see pp. 6 and 12, and Figs 1.8 and 1.15). It is easy to anticipate what happens in easy cases by imagining the spot position in slow motion. The simplest situation is that in which the two a.c. amplitudes are the same, and in which the voltages are in phase. The electron beam then traces out points that always have the same displacements in vertical and horizontal directions, so that a diagonal line of slope 45 degrees appears (Fig. 2.15(a)). If the two voltages are 180 degrees out of phase, the Y plates deflect the spot downwards as the X plates deflect it to the right, so that a straight line is drawn out, but this time with the 45-degree slope in the reverse sense (Fig. 2.15(b)). Now let us consider what happens when the voltage applied to the Y plates is in advance of (leads) the voltage applied to the X plates by 90 degrees. At the moment that the a.c. voltage on the X plates is about to start its positive half cycle, the voltage on the Y plates is already at its maximum and is about to start decreasing. The spot on the screen therefore starts in the horizontal centre position but at the maximum of its vertical displacement, and as the horizontal displacement increases, the vertical displacement decreases. By the time the spot is at the extreme of its right-hand motion, it has also reached the centre of the vertical position, and is about to start being depressed below the centre. Following this kind of reasoning through a complete cycle soon shows that the complete motion is a circle drawn out in the clockwise direction (Fig. 2.15(c)). If the frequency is above roughly 10 Hz, the human eye will not be able to see the direction of motion, but will still observe a circle on the screen. It is obvious that if the Y-plate voltage is behind (lags) the X-plate one by 90 degrees (or leads by 360 − 90 = 270 degrees), a circle will again appear, but the direction

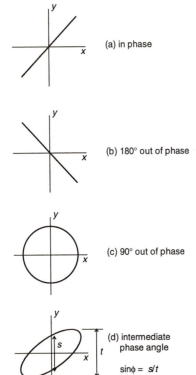

(a) in phase

(b) 180° out of phase

(c) 90° out of phase

(d) intermediate phase angle

$\sin\phi = s/t$

Fig. 2.15 Lissajous figures produced by an oscilloscope used in X–Y mode for identical input frequencies but different phase relationships

of rotation will, in principle, be counterclockwise. For any intermediate phase angle, an ellipse will be seen, and the phase angle between the two voltages can be calculated simply from the dimensions of the ellipse, as shown in Fig. 2.15. Thus, if some device is used to vary the phase angle φ between two a.c. signals of identical frequency, then as φ is increased from zero degrees, the straight line of Fig. 2.15(a) opens up to become an ellipse, and then a circle at 90 degrees, collapsing again to an ellipse with increasing φ, until the straight line of Fig. 2.15(b) is obtained at 180 degrees. The reverse sequence of changes occurs for φ between 180 and 360 degrees, with the pattern being the same for an angle 360 − φ as it is for an angle φ (but with the opposite direction of motion of the spot).

Stationary patterns on the screen like those shown in Fig. 2.15 can only be obtained if the frequencies of the two voltages applied to X and Y plates are exactly the same. If there is even a small difference, the phase relation between the two will change with time, and the patterns will vary continuously over the entire sequence just described at the end of the last paragraph. This result affords a method of *frequency comparison*. The number of times that the sequence of patterns repeats itself in one second is equivalent to the frequency difference between the two a.c. voltages. Since it would be fairly easy to observe the sequence over a period of one hundred seconds or more, it is apparent that the difference in frequencies can be determined to better than one-hundredth of a hertz, or that one source can be tuned to another within that accuracy. Given that the oscilloscope might be examining a.c. signals of frequencies of 100 MHz or more, these numbers would represent a precision of one part in 10^{10}, which is hard to match in anything but the most sophisticated digital frequency meters, and yet is achieved by a simple and familiar piece of laboratory equipment that presents the information in a way immediately assimilated by the human eye.

So far, we have confined ourselves to the patterns obtained when the frequencies of the two a.c. signals are identical. It is only a small step to work out the patterns for other simple frequency relationships. Figure 2.16 shows the results for a vertical:horizontal frequency of (a) 2:1 and (b) 1:3. Ratios such as 2:5 and 3:2 also give quite easily distinguished patterns, but once the ratios require larger integers for their description, the number of crossing lobes in the pattern increases to such an extent that only a blur is seen. At frequency ratios *just* off the small integer ratios, a 'revolving' pattern is seen of the same nature as that seen for nearly equal frequencies. The use of the oscilloscope in the X–Y mode thus allows simple frequency ratios to be determined or permits frequencies to be matched to set ratios within very close tolerances.

2.6 Potentiometers and bridges: null methods

Null methods

Underlying all null methods is the possibility of balancing one voltage or current against another in order to get a net zero output. The idea is shown

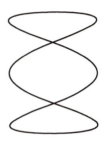

(a) vertical:horizontal = 2:1

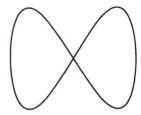

(b) vertical:horizontal = 1:3

Fig. 2.16 Lissajous figures for different input frequencies that bear a simple integral numerical relation to each other

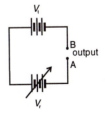

Fig. 2.17 The balancing of opposed test and reference voltages When the voltages are identical the output is zero

schematically in Fig. 2.17 for the balancing of two voltages. Voltage V_t is the voltage to be tested, and V_r is a reference voltage whose value can be varied. If V_r is equal to V_t, there is no potential difference across the output terminals, a condition that can be tested either with a high resistance voltmeter or a (relatively low resistance) current measuring device.

Null methods play a conceptually most significant part in electronic instrumentation, and, although for some applications they have become superseded by newer electronic devices, they continue also to be of great practical importance. There are several reasons why null methods are potentially so useful. The first is that the measurement of output current or voltage does not have to be accurate or even performed with a calibrated instrument. The only requirement is that the instrument be capable of detecting the balance point with the desired sensitivity. Accuracy in the determination depends on the accuracy with which the reference quantity is known. Comparison of the known and unknown quantities is the essential feature. Exact analogies can be found with the most basic scientific measurements such as weighing. Given adequate delicacy of the suspension, a null-reading two-pan balance has an accuracy of measurement that rests on the accuracy with which the mass of the counterweight is known. In contrast, a spring balance needs to be calibrated in advance (and the calibration may alter with factors such as the local acceleration due to gravity). For many years, electrical null points were determined with various forms of *galvanometer*, often using a mirror attached to a suspended moving coil movement in order to deflect a light beam. High sensitivities could be obtained, but the absolute sensitivity was often unknown and of no importance if the instrument was to be used only to detect a balance point. Null points are now usually measured with more convenient detectors, although galvanometers from a previous era are quite often to be found lurking within sophisticated and modern pieces of chemical instrumentation. Whatever balance detector is used, the principle remains the same, and the relative unimportance of absolute calibration brings with it great benefits.

A second advantage of the null method of voltage comparison is immediately apparent if it is remembered that zero current flows at the balance point. No load is imposed on the test voltage source, so that the true e.m.f. is measured (cf. p. 23), so long as the balance point is found exactly.

The potential divider

The resistive *potential divider* provides an important way in which known voltages can be can be derived from a source of higher voltage. Figure 2.18 illustrates a two-element divider. Since a common current flows through each resistor, the voltages across each are in the ratio of the resistances themselves, and it is therefore a trivial task to calculate these voltages in terms of the applied source voltage. Some quantitative voltage relations are shown in the text accompanying the figure. The accuracy with which the relative voltages are known depends on the accuracy with which the resistances are known; the source voltage must be known accurately to obtain accurate absolute output

Fig. 2.18. A two-element potential divider. Some simple formulas give the voltages across each resistor:

$$V_1/V_2 = R_1/R_2$$
$$V_0 = V_1 + V_2$$
$$V_1 = V_0 R_1/(R_1 + R_2)$$
$$V_2 = V_0 R_2/(R_1 + R_2)$$

voltages. Precision resistors are relatively easy to fabricate from measured lengths of wire, and laser trimming of deposited conductive films also plays its part in modern technology. A continuously variable voltage source can be made by tapping off a potential from a length of resistance wire (which itself may be stretched out straight or wrapped in a spiral), as indicated in Fig. 2.19. Two points are worth making about the use of a potential divider. To retain the inherent accuracy of the divider, whatever is connected to it must not alter appreciably the resistance across which it is connected. In other words, the load resistance must be high compared with the dividing element, and the load must draw negligible current. This condition can be often met in practice, especially in the two applications that are to be discussed in the next sections. The second point is that the load presented by the divider resistance itself (R_1 + R_2) to the voltage source must be considered in designing the circuit and interpreting its behaviour.

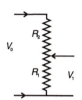

Fig. 2.19 The fixed-resistor circuit of Fig. 2.18 has been modified to provide a continuously variable voltage source by using a slide-wire or tapped resistor

The potentiometer

The *potentiometer* is a familiar piece of laboratory equipment that combines the features of a null voltage measurement (Fig. 2.17) with a potential divider network (Figs 2.18 and 2.19) to generate a known reference voltage. Figure 2.20 shows the variable reference source of Fig. 2.17 replaced on the left-hand side by the circuit of Fig. 2.19. A combination of switched fixed resistors and sliding variable resistors is often used in the divider. The balance point, obtained by adjusting the divider, is determined with a meter of suitable sensitivity. Laboratory potentiometers are often calibrated by determining the resistance ratio needed to balance the voltage of a *standard cell*, which is an electrochemical cell of closely specified construction and chemical composition, whose load-free potential (i.e. its e.m.f.) is known accurately at a specified temperature. As pointed out earlier, a correctly balanced potentiometer does not load either the test source or the standard cell.

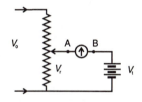

Fig. 2.20 The potentiometer. The reference voltage source of Fig. 2.17 has been replaced by the variable-voltage source of Fig. 2.19

Bridges

From the potentiometer, it is only a small step to the *Wheatstone bridge*. Figure 2.21 shows the circuit in a rather unusual form in order to emphasize the connection between the bridge and the potentiometer. In the bridge, the test source of voltage of Fig. 2.20 is replaced by a second potential divider energized by the same voltage source as the left-hand divider. No current flows through the meter, and the bridge is thus balanced, when $V_1 = V_3$. The relationship between the four resistors for balance can be expressed in several ways, of which the form $(R_1/R_2) = (R_3/R_4)$ is one of the most useful. In Fig. 2.22, the bridge is drawn in the conventional way: the circuit is unchanged.

The application of the Wheatstone bridge is in determining unknown resistance values. For this purpose, the unknown resistance constitutes one 'arm' of the bridge, such as the R_1 position, and the other three resistors altered to achieve the balance. Usually two of these resistors would be left

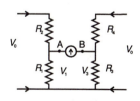

Fig. 2.21 The Wheatstone bridge, drawn to illustrate the use of two opposed and balanced potential dividers of the kind shown in Fig. 2.18

Fig. 2.22 The Wheatstone bridge circuit of Fig. 2.21 drawn in a more conventional way

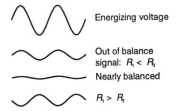

Energizing voltage

Out of balance signal: $R_1 < R_3$
Nearly balanced

$R_1 > R_3$

Fig. 2.23 Output waveforms from an a.c. resistive bridge (voltages at point B with respect to A)

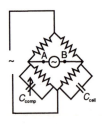

Fig. 2.24 Balancing of stray capacitance with a compensating capacitor in an opposite arm of the a.c. bridge

fixed and only the third altered. For example, R_2 might be chosen to equal to R_4, and R_3 adjusted. The value of R_3 at the balance point is then directly equal to that of the unknown R_1.

For some purposes in the laboratory, it is desirable to energize the bridge with a.c. rather than d.c., but so long as all four arms behave as pure resistances (that is, current and voltage are in phase at all times), all the arguments advanced so far remain valid at each point of the a.c. cycle. Balance points in audiofrequency a.c. bridges are sometimes obtained using headphones to detect audibly an off-balance situation. The absence of an absolute sensitivity calibration for the headphones is no disadvantage because the bridge balance is a null point. However, other methods of detection are often used, and some of these rely implicitly on the phase relations between input and output signals. Just as in the d.c. case, the output is positive on one side of the balance point and negative on the other side of it, so with the a.c. bridge the output waveform is in phase with energizing waveform in one direction of off-balance and reverses phase by 180 degrees on the other side of the balance, as illustrated in Fig. 2.23.

The situation becomes more complicated if there is capacitance (or inductance) associated with the resistive elements. 'Stray' capacitance is usually present (for example, in cells used to determine the conductivity of solutions, as described on the next page). The capacitative reactance (see pp. 12–13) produces currents and thus voltages in the bridge that are 90 degrees shifted in phase from the energizing waveform, and can thus never be balanced by a resistance in which currents and voltages are exactly in phase. An easy way of dealing with the problem is to compensate the out-of-phase voltages with a small variable capacitance (it could just be a pair of twisted, but insulated, wires), as indicated in Fig. 2.24. The voltages at the 'A' and 'B' ends of the two capacitors now have the correct phase relationships with each other to be able to balance independently of the purely resistive part of the circuit. Several methods exist for finding the balance of resistive and capacitative elements separately, including the use of dual-beam oscilloscopes to examine the phase relationships between output and input waveforms. A more subtle method uses the Lissajous figures produced by the oscilloscope in X–Y mode (see pp. 28–29). The horizontal plates are connected to the bridge-energizing supply and the vertical plates to the bridge output. Off balance, an ellipse will be seen in general (Fig. 2.15(b)). As the compensating capacitance is adjusted towards the correct value, the ellipse should collapse to a straight line (Figs 2.15(a) and 2.15(b)). The variable resistor can now be adjusted to balance the in-phase voltages and flatten the diagonal line to become a horizontal one at balance.

The use of the compensating capacitor just described is really an extension of the use of the a.c. bridge to measure capacitances and inductances. Figure 2.25 shows a capacitance bridge. In this circuit, there are no resistors at all in the lower two arms. The two upper resistors allow currents to flow in the capacitors (it leads the voltage by 90 degrees, as discussed on p. 13). If the capacitances, and thus the reactances, are equal, the voltages at points

'A' and 'B' will be identical at all parts of the a.c. cycle, even though they are both leading the applied voltage by 90 degrees. The bridge is therefore balanced. The balance point can be found using any of the a.c. detectors discussed previously, and an unknown capacitor may be balanced against a known one to determine its capacitance. Similar bridges constructed with two inductors in place of the capacitors are used to measure unknown inductances.

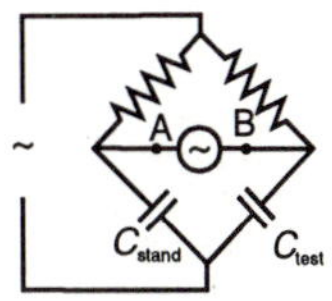

Fig. 2.25 The capacitance bridge

Solution conductivity measurements in chemistry

One important use of the a.c. bridge in chemistry is in determining the conductivity of ionic solutions. Such measurements give access to a variety of important quantities, including the mobilities and concentrations of ions, that can be used ultimately to obtain information about chemical equilibria in the solution and the solvation of the ions. In these determinations, a cell is used in which two platinum plates are immersed in the solution under test (Fig. 2.26), and the resistance measured. The conductance is the inverse of the resistance. The specific conductance (the conductance across opposite faces of a cube with sides of unit length: see Fig. 1.1) can, in principle, be calculated from the known dimensions and separations of the plate, although it is more usual to calibrate the cell with a test solution of known specific conductance. Just because the conduction is due to the movement of ions from one electrode to another, it follows that ions of positive charge will move towards the negative plate, and ions of negative charge will migrate towards the positive plate. If this separation of charged ions is allowed to continue, the electrical and chemical properties of the solution may be altered: the solution will become *polarized*, and it may even be *electrolysed*. One way of avoiding these problems is to make the resistance measurements using a.c. rather than d.c. voltages. Since the bridge is the obvious way of obtaining accurate resistance determinations, the a.c. bridge is of particular value in electrochemical conductivity measurements. However, it should be recognized that the the two parallel plates constitute a small capacitor in their own right. The problem about the out-of-phase current raised on the last page is thus inevitably present. A compensating capacitor, and the appropriate null detection techniques described earlier, can be used to offset the cell capacitance. A neater trick is to replace the compensating capacitor with a dummy cell, connected in a similar way to the bridge as the main cell, and identical to it in all respects except that it contains pure solvent rather than the solution under investigation. The capacitances of the two cells and their associated wiring should then be virtually the same. If they are not, a very small 'trimming' capacitance can be added to one arm or the other of the bridge. The configuration has the further advantage that the dummy cell also provides compensation for the conductivity of the solvent, so that only the component resulting from the solute ions is measured.

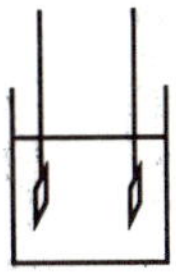

Fig. 2.26 A simple conductivity cell

3 Input transducers

3.1 Conversion into electrical phenomena

In this chapter, we examine some of the most important ways in which non-electrical variables that are measured in experiments may be converted into electrical phenomena for processing and conditioning of the signals. Our attention here is initially focused on the measurement of the most basic physical quantities such as spatial position, temperature, or light intensity. We shall see subsequently how one *non*-electrical phenomenon may be converted to another *non*-electrical one that is ultimately interfaced to an electrical transducer. A typical example might be in the measurement of pressure; the pressure of a gas might be used to distort a capsule in a type of barometer, and the displacement in position of some portion of the capsule translated into an electrical signal that is processed. Discussion of electrochemical measurements is deferred entirely until the end of the chapter, because the electrical signal is produced directly by the chemical phenomenon of interest, and the input transducer is the electrode or similar device that probes the chemical system.

As explained in the last chapter, electrical signals may be in analogue or digital form, and the type of signal may be determined from the outset by the input transducer. Some devices can produce either analogue or digital output, as we shall see in connection with the detection of charged particles (Section 3.4) and light (Section 3.6). Other important operations are clearly digital, such as the counting of the number of objects that pass a given point by interrupting a light beam that itself produces an electrical response. In most cases, the electrical signal treatment and output can be analogue or digital according to ease of processing and convenience of display, because it is usually easy to interconvert the different types of signal (see p. 22).

Transfer functions and linearity

Regardless of the particular non-electrical (say α) and electrical quantities (say β) that are involved in the input step, it is essential to know how β depends on α if the final output is to be a quantitative measure of the phenomenon under investigation. This relation is the *transfer function* of the input transducer, and is a form of calibration of the electrical output in response to the input quantity. The same is true, of course, of the output transducer. It is most convenient if the transfer function is linear, that is if β is proportional to α, because then it is easier both to know how most effectively to calibrate, and to interpolate within the calibrated range, or extrapolate outside it. Other simple mathematical functions, such as logarithmic, exponential, or square-law dependences, are almost as convenient, especially since elements in the electrical conditioning chain can be fashioned with the inverse response and thus produce a linear output. The incorporation of computing power within the signal conditioning chain extends the range of acceptable transfer behaviour of the input transducer, and, if all else fails, the computer can be equipped with a 'look-up table' that contains calibration factors for all possible input values. The fact remains that this look-up table has to be constructed, and, unless an analytical mathematical form is available for the transfer function, it has to be obtained from a point-by-point calibration procedure in which the scope for interpolation is limited. There thus remain considerable advantages in using devices that are capable of exhibiting linear transfer behaviour, and in selecting conditions where that behaviour is followed.

Types of transducer

Before moving on to an examination of the ways in which the principal physical phenomena can be converted to electrical signals, it is useful to consider the different types of the electrical response that can be generated. In this section we examine only the general categories, with a brief mention of typical examples that are then treated in more detail in later sections.

First, the input stress may cause an alteration in the 'passive' electrical characteristics of a circuit element such as a resistor, capacitor, inductor, or transformer. Measurements of the appropriate electrical quantity, such as resistance or capacitance, then reveal information about the input stress. Typical examples might be the change of resistance of a wire with changing temperature, which gives access to electrical temperature measurements, and the change in capacitance between two plates as one moves relative to the other, which permits determination of position and movement.

Secondly, the input phenomenon may generate electrical energy. Obviously, the non-electrical phenomenon must also be a source of energy such as motion, heat, light, or even chemical energy. A motion transducer might consist of a magnet inserted into a coil of wire attached to the system that moves. While the magnet and coil move relative to each other, lines of force are cut, and a voltage is induced in the coil. We shall meet examples of the conversion of heat and light into electrical energy later in this chapter, and chemical energy is clearly involved in the electrochemical generation of e.m.f. to be discussed in Section 3.9. Optimum linearity of the transfer function of all these devices is often obtained if they are operated as voltage sources (without a load) or as current sources (short circuited).

The third type of transducer to be examined here generates electrical charge (strictly speaking, negative and positive charges are separated), and the electrical charge is determined by the current flow that it permits when an appropriate electrical potential difference is applied to the system. The vacuum photocell is an example of this kind of transducer. Light of suitable wavelength ejects electrons from metals (the *photoelectric effect*), and these electrons may be detected by arranging for a positively charged anode to pick them up and thus cause a current to flow in a circuit completed by connection

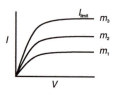

Fig. 3.1 Current–voltage curves for typical transducers that generate charge. Beyond a certain voltage, all the charge is collected, and the current approaches a constant limiting value, I_{limit}

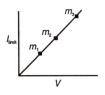

Fig. 3.2 The limiting currents shown in Fig. 3.1 depend in a linear manner on the magnitude m of the phenomenon generating the charge

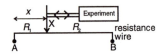

$R_1 \propto x$ for uniform cross section wire

Fig. 3.3. A resistance transducer for position. The arrows indicate the directions of motion

to the metal cathode from which the electrons were ejected (see Section 3.6) Devices of this kind show rather unusual current–voltage relations, since the current is limited by the supply of charge rather than by the voltage across the circuit. These *constant current* sources thus disobey Ohm's law completely Figure 3.1 shows a typical set of current–voltage curves for differen magnitudes, m, of the phenomenon generating the charge: m would, fo example, be the light intensity in the case of the photoelectric cell jus mentioned. Beyond a certain voltage, which corresponds more or less to tha needed to ensure that all the charge carriers are collected, the current reaches a constant limiting value, i_{limit}, for a given value of m. Figure 3.2 then shows how this limiting current varies with m. If one is fortunate, and the number of charge carriers is directly proportional to m, the relationship is linear. This is the situation represented in the figure, and it does apply, in reality, to devices such as the simple photocell.

3.2 Position, displacement, and strain transducers

In most applications, it is the displacement from an initial position that is measured, the absolute position being inferred from the known starting point In this section we consider linear displacements, but almost every technique can be adapted to measuring angular displacements.

Resistive transducers

One of the simplest ways of measuring a displacement is by attaching the moving object to a slider that makes contact with a resistance wire, as illustrated schematically in Fig. 3.3; the adaptation for measuring angula displacements with a curved wire is obvious. With a wire of uniform cross-sectional area, the change in resistance between the slider and one end of the wire varies linearly with the change in position of the slider. The resistance itself can be measured by any of the standard techniques, one of the most suitable being to incorporate R_1 and R_2 of Fig. 3.3 into the arms of a Wheatstone bridge (see pp. 31–32 and Fig. 2.22). Alternatively, a known potential can be applied to the ends, A and B, of the wire, and the voltage tapped off at X measured. The device is now a potential divider (see pp. 30–31 and Fig. 2.19), and the voltage between A and X will be a linea function of the displacement x. A complete measuring instrument might then consist of this set-up connected to a digital voltmeter (see p. 25). Suppose the wire to be one metre long and connected to a source of one volt. The reading in volts on the digital voltmeter would then be equal to the displacement o the slider, x, in metres.

Capacitative and inductive transducers

A drawback to the use of the resistive displacement transducer is that it is bound to impose a mechanical load on the moving system. For heavy-duty systems, where the forces involved are large, this load may be of negligible importance, but in delicate experiments it may seriously interfere with the measurement. Recourse then has to be made to methods where there are no mechanical forces imposed by the electrical system. Capacitors, inductors, and transformers can be used in various ways to achieve such measurements. Figure 3.4 shows how the scheme might work for a capacitance transducer in which one plate is attached to the moving system and the other is kept fixed. As the gap between the plates narrows, the capacitance increases. Figure 3.5 shows a similar arrangement with an inductor. Here, a 'slug' of high permeability material that is inserted into the coil is attached to the moving system, and as its position within the coil changes, so the inductance alters (see p. 13). One convenient way of determining capacitances or inductances is to use an a.c. bridge (see pp. 31-32). Indeed, the variable capacitance or inductance system can form two arms of a bridge, as illustrated for the inductance case in Fig. 3.6 (compare with Fig. 2.25). The slug is arranged so that, as the inductance of one coil increases, the other decreases. The arrangement is exactly equivalent to the bridge measurement of the resistances just described in connection with Fig. 3.3. One great advantage of these bridge connections is that spurious changes in resistance, capacitance, or inductance resulting from outside influences, such as changes in ambient temperature, tend to be cancelled out, leaving only the effect of interest to alter the balance of the bridge. It might be noted here that, although it is usual to balance bridges in absolute measurements of resistance, inductance or capacitance, it may be more useful in the present applications to determine the offsets from balance by measuring the out-of-balance signal at the detector as a voltage.

Another method for measuring inductance or capacitance that is sometimes employed for displacement transducers involves making them part of a resonant tuned circuit (see p. 15 and Fig. 1.21) that determines the frequency of an oscillator (see p. 18). The frequency generated will thus depend on the capacitance or inductance of the transducer, and gives a measure of the displacement. One can envisage an instrument that uses a digital frequency meter (see p. 25), with the counters arranged so that the output is a direct display in units of length.

Conceptually rather similar to the inductance bridge, the *differential transformer* provides another way of measuring displacement. Figure 3.7 shows a primary winding, energized at R and S by a source of a.c. voltage, sitting between a pair of secondary windings connected in *antiphase*. This reverse-phase connection means that, if everything about the transformer is truly symmetrical, all voltages induced in one secondary coil are exactly cancelled by those in the other, and there is zero output across the terminals A and B. If a slug of high permeability material is inserted into the core, this balance is upset, and a voltage now appears at the output terminals; the magnitude depends on the displacement from the symmetrical position (and the phase depends on which side of this position the displacement occurs).

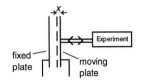

Fig. 3.4 A capacitance displacement transducer

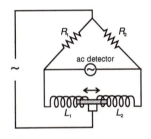

Fig. 3.5 An inductance displacement transducer

Fig. 3.6 A differential inductance transducer arranged in two arms of an a.c. bridge

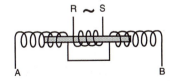

Fig. 3.7 The differential transformer used as a displacement transducer

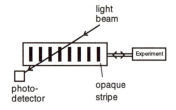

Fig. 3.8 An optical position transducer

Fig. 3.9 Moiré fringes produced by two identical striped patterns, one of which is tilted at a small angle θ with respect to the other

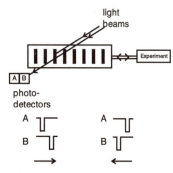

Fig. 3.10 Use of a pair of photodetectors to determine the direction of motion of the plate in Fig. 3.8. If the interruption in beam A precedes that in beam B, the plate is moving from left to right, and vice versa

Optical methods

All the methods of measuring displacement presented up to now produce an output that is initially analogue, although it may be converted to a digital form later on. Optical techniques provide one way of obtaining a digital result from the outset. Consider the arrangement outlined in Fig. 3.8. A light beam passes through a piece of transparent material attached to the moving system, and is then incident on a photodetector (see Section 3.6). On the transparent material are marked opaque stripes, so that as the system moves, the light beam is interrupted as each stripe passes in front of it. By counting electrically the number of times that the beam has been broken, and knowing the separation of the stripes, the extent of movement can be determined immediately and in digital form. It is worth noting, however, that the resolution of the device is determined by how close together the stripes are marked, and that in turn may be limited by the width of the light beam. A variant of this method overlaps two pieces of material on which close-spaced regular patterns are marked and that produce Moiré fringes. These fringes shift by large amounts for small relative displacements of the patterned material, and can be counted easily by the optical–photoelectric method (Fig. 3.9). Of course, the arrangement shown in Fig. 3.8 only provides information about the distance travelled, and not about the direction of motion. A small modification allows the additional information to be obtained. If a second beam and detector are set up so that the two beams are closer together than the stripes on the moving plate, then the order in which the interruptions occurs reveals the direction of motion, as illustrated in Fig. 3.10.

Strain gauges

The devices discussed so far in this section do not, in principle, impose any significant load on the system being investigated, although there will, in practice, be frictional losses with the sliding-contact resistance transducer. Rather different in approach are *strain gauges*, which measure the distortions that follow mechanical displacement. If a piece of resistive material is distorted, the changes in length and cross-sectional area lead to changes in resistance, and the resistivity itself may alter. Determination of the resistance thus provides a measure of the strain to which the material is subjected. Practical strain gauges may be made up of fine resistance wire, or of patterns of thin metal films deposited on a flexible insulating substrate material. As usual, bridge methods are called for in order to eliminate the effects of temperature changes on resistance, and the deposited strain gauges are usually fashioned with pairs of balanced elements so that distortion in one direction increases the resistance of one element and decreases that of the other. Temperature changes alter both resistances equally, and are thus nulled out in the bridge.

Certain crystals are *piezoelectric*: that is, they generate an electric field when they are distorted. The voltage induced can therefore be used as a measure of the strain in the crystal and of the displacement that produced it. We note in passing that a mechanical force is needed to yield the potential, and that if electrical work is to be done by the piezoelectric device, energy will be taken from the mechanical system. The piezoelectric effect is used in several types of pressure transducer, as well as in microphones and other kinds of motion and force detector.

3.3 Velocity transducers

The measurement of velocity involves determining the rate of change of position with time, so that any suitable position transducers can be used in combination with a timing device to calculate the velocity. Mathematically, the velocity is the differential with respect to time of the position, and the differentiation can be achieved electronically, either as an analogue or a digital operation. Several simple electronic circuits differentiate the input signal. For example, the very simple *R–C* circuit of Fig. 1.19(b) performs the operation. Digital signals can be handled by using a ratemeter instead of a simple counter: that is, the number of counts accumulated in a fixed sampling interval is displayed rather than the total number of counts. One very important application in chemical experiments is to measure the angular velocity — the rate of rotation — of some device or other. The optical position transducer of Fig. 3.8 can be harnessed for this purpose, as indicated in Fig. 3.11. A disk divided into sectors with alternate portions opaque or transparent is attached to the rotating axis, and the output from the photodetector is a multiple (the number of pairs of sectors) of the frequency of rotation. We shall meet derivatives of this *chopper* disk or wheel in Chapter 5 (see p. 69).

With a moving, rather than a static, system, it is also possible to generate electrical energy by changing the position of a magnet inside a coil. As the magnetic field changes, a voltage is generated in the coil, a situation illustrated in Fig. 3.12. Either the coil or the magnet may be moved, and the magnet may be a permanent one or an electromagnet. The voltage induced is proportional to the rate at which the magnetic field changes and is thus dependent on the velocity of motion. Similar methods, using rotating magnets, can be used to measure angular velocities: the transducer is then a kind of alternator, with the output frequency equal to the rate of rotation, and the induced voltage increasing with increasing angular velocity.

3.4 Detection of charged particles in the gas phase

The charged particles that are encountered in gas-phase chemistry are electrons, and positive and negative ions. Often, the ionic species are found in mass spectrometry, but there are a number of other ionization phenomena, including photoionization, the photoelectric effect, and ionization in flames that are of importance to the chemist, and we shall meet several of them later. In this section, we consider the general principles of their detection. Collision

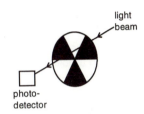

Fig. 3.11 A chopper wheel used in the optical detection of rotation. This device is the angular analogue of the linear motion transducer shown in Fig. 3.8

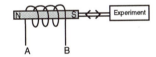

Fig. 3.12 Inductive motion transducer. In this version, a permanent magnet is moved with respect to a fixed coil, in which a voltage is generated

of any of the charged species with neutral gas molecules can lead to secondary ionization, so that where the number (and identity) of the species contains the information of interest, it is usual to operate at gas pressures sufficiently low that collisions are infrequent enough not to invalidate the results. Pressures of 10^{-4} torr (roughly 10^{-7} atmospheres) or less are usually acceptably low, although more demanding experiments may require yet lower pressures. Other experiments again, such as those with flames, are performed at atmospheric pressure.

Charge collectors

It is evident that electrical methods would be the obvious candidates for the study of charged species. The very simplest way of determining the concentration of charged particles is to measure the current that flows when the charge carrier is attracted to a plate or cup of opposite polarity, as illustrated for electron detection in Fig. 3.13. It should be observed that the circuit must be completed by providing a return path to the source of charge carriers. When the applied voltage is sufficient to attract virtually all the charge carriers, the current reaches a limiting value, in the way shown in Figure 3.1, which is determined by the rate of supply of the charged particles. The limiting current is a function of this rate of supply (Fig. 3.2).

Electron multipliers

A current of one amp is produced by a flow of 6.2×10^{18} electrons per second; put another way, an electron flux of 10^7 electrons per second (quite a high value in many experiments) will support a current flow of about 1.6 picoamps, which is beginning to get difficult to measure by direct collection techniques. A way around this problem is to use an *electron multiplier*. Figure 3.14 shows the principle of the electron multiplier. The electrons to be detected are separated from the collecting anode by several (perhaps ten or more) *dynodes*, which in the structures illustrated are made up of slotted metal. Each dynode is held at a successively higher positive potential by means of a suitable chain of resistors acting as a potential divider. If an electron is accelerated sufficiently, when it hits a suitable surface it can eject several secondary electrons. The metal dynodes are usually coated with some efficient secondary emitting material. At each dynode, each electron is thus multiplied by some factor. One electron might, for example, eject six from each dynode, so that with n dynodes, the number of electrons leaving the last dynode for the anode would be 6^n for each primary electron that reached the first dynode. For a multiplier with eight dynodes, the electron *gain* might thus be about 1.6 million, and much higher gains are available with multipliers possessing more dynodes. With our initial example of 10^7 primary electrons per second, the final anode current would now be roughly 2.6 microamps, which is relatively easy to measure. The great beauty of the electron multiplier is that secondary electrons are, at least in principle, ejected only if a primary electron is present. Thus, although the multiplier is a kind of

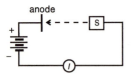

S = electron source (cathode)

Fig. 3.13 Collection of charge carriers with a polarized plate. In this example, electrons from the source S are collected by a positively charged anode. For positively charged species, such as positive ions, a negative collector must be used

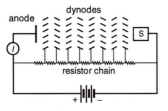

S = electron source (cathode)

Fig. 3.14 One form of electron multiplier. The whole assembly of source, dynodes, and anode normally operates in a near-vacuum environment

amplifier, it is one that amplifies without introducing its own noise into the signal, a matter of some importance as we shall see in Chapter 5. Of course, the situation is not quite as simple as that outlined here, because not every electron is multiplied by the same factor, and some electrons may be produced from the dynodes thermally, or possibly by the arrival of cosmic rays and by radioactive decay events taking place in minor isotopes of the materials of which the multiplier is constructed. Nevertheless, the electron multiplier has very great advantages over other methods for the detections of electrons. It can even be used to measure the flux of ions. For example, if the first dynode is made highly *negative* with respect to a source of *positive ions*, they will be accelerated into that dynode and eject secondary *electrons*. If the following dynodes and the anode are successively less negative (that is, more positive), then electron multiplication and detection will occur.

A modification of the electron multiplier that has found increasing application where robustness and simplicity of operation are required is the *channel electron multiplier* (often shortened to *channeltron*, a term that is properly used as the tradename of one particular manufacturer). The device consists of a small bore tube of glass or other insulator, the surface of which has been made partially conducting by incorporating some suitable material such as lead and lead oxide. A typical construction is shown in Fig. 3.15. Now the tube acts as its own potential divider when the cone end (cathode) is made negative and the collector (anode) positive. Accelerated electrons make collisions with the walls in the 'channel', and, with the right choice of surface, secondary electrons are ejected. The process is repeated many times down the tube, with the result that high electron gains can be achieved. The curved or spiral shape of many of these multipliers is provided to reduce interference from positive ions generated by electron impact within the channel on residual gases in the system. *Microchannel plates* consist of hundreds or thousands of parallel small multiplying channels constructed within the same plate as a kind of honeycomb structure. They find application in investigations of the two-dimensional distributions of electrons and ions, and are thus valuable in various kinds of imaging system.

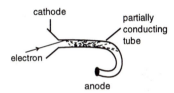

Fig. 3.15 A channel electron multiplier showing schematically the first two steps of electron multiplication (dashed lines)

Electron counting

Since electrons (and ions) are individual discrete charged particles, the possibility of detecting the arrival of individual particles at a charged electrode and counting them directly suggests itself as an alternative to inferring the numbers of charged particles from the current flow. Where the rate of arrival or production of charged particles is rather small, this technique appears to have much to commend it. The drawback is the small charge on individual electrons, of about 1.6×10^{-19} coulombs, which is impossibly small for measurement. However, if the individual electrons are first subject to the multiplication process just described, the bursts of charge are easily detected, and *pulse counting* techniques can be used to determine absolute numbers or rates of arrival corresponding to individual electron events down to one pulse per second or less.

The flame ionization detector

Since ionization in flames at atmospheric pressure was alluded to earlier, this seems an appropriate place to introduce one widespread use of the phenomenon. When most organic compounds are burned in air, they generate both positive ions and electrons. The *flame ionization detector (FID)* employs this phenomenon to detect and quantify the presence of ionizable species in the eluent gases from a gas chromatograph. Figure 3.16 shows the detector in simplified form. The gases coming off the GC column are mixed with hydrogen (if hydrogen is not the carrier gas), and burned in a small flame supported on a metal burner. Surrounding the flame is a collector electrode and a circuit is completed by providing a return path to the burner. With the collector charged negatively with respect to the burner, any positive ions present are attracted to it and carry a current. Electrons carry negative charge in the reverse direction. A current thus flows in the circuit, and the magnitude is nearly proportional to the concentration of organic material for the tiny amounts of material present. Recording of this current as a function of time as different materials come off the column thus provides the gas chromatogram itself.

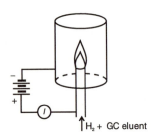

Fig. 3.16 A simplified flame ionization detector

3.5 Measurement of temperature

Resistance measurements form the basis of certain types of temperature transducer. Good conductors, such as metals, exhibit an increase in resistance with temperature. Over fairly limited temperature ranges, the resistance is often roughly a linear function of temperature. Platinum is the metal most commonly employed — in the *platinum resistance thermometer* — and at room temperature its resistance increases by about 0.3 per cent per degree centigrade. Bridge methods of measuring resistance are often employed in accurate work. However, for some purposes, the changes in resistance of pure metals are insufficient, and in these cases *thermistors* made from semiconductors can be used. Pure (*intrinsic*) semiconductors conduct as a result of the promotion of valence electrons to the conduction band. The excitation may occur thermally, and the population of charge carriers will follow a distribution of the Boltzmann type. As a result, the resistance of such semiconductors drops rapidly with increasing temperature. Thermistors can thus offer high sensitivity to changes in temperature. The transducing element can also be made physically very small and of low thermal capacity so that it can probe highly localized zones. The one drawback is that different samples of 'pure' semiconductor may in reality possess differing levels of purity and thus of electrical behaviour. The reproducibility of absolute temperature measurements with different thermistors may not, therefore, be as good as with the platinum resistance thermometer. Thermistors also suffer from a relatively low upper-temperature limit (about 350°C).

All methods of determining resistance ultimately depend on determining the voltage drop across the resistor when a known current passes through it (see pp. 23–24 and 31). With either metallic resistance thermometers or thermistors, it is essential that the heating effect of the test current produce a negligible temperature rise in the probe, or the temperature measurement will be invalidated.

Heat energy can also be used to generate electrical energy, and the voltage or current produced used to determine the temperature. The basic device that performs the energy conversion is the *thermocouple*. If two dissimilar metals, M_X and M_Y, are connected to form a pair of junctions as shown in Fig. 3.17, and the temperatures of the two junctions, T_H and T_C, are different, then a voltage will be generated across the terminals A and B. The output voltage can be expressed as a polynomial in temperature, as indicated alongside the figure. For many metal pairs, the higher order terms are swamped by the first one, so that the voltage is nearly proportional to the temperature difference $T_H - T_C$, at least over a limited temperature range. Calibration tables are available for the most common pairs of metals. One of the junctions is held at a constant reference temperature, for example by maintaining it in a bath of melting ice. A typical metal pair such as iron and the alloy constantan produces an output of just over 5 millivolts for a temperature difference of 100°C. These relatively small voltages mean that amplification is almost always needed if the output is to be displayed or recorded.

In some applications, many hundreds of small thermocouple pairs physically laid out next to each other may be wired in series as illustrated in Fig. 3.18 to make a *thermopile*. The output voltage from the thermopile is, of course, the output of a single device multiplied by the number of junction pairs in the pile, and is much higher than that available from the simple thermocouple. Thermopiles have been used for the detection and measurement of *radiant energy* (infrared, visible, and even ultraviolet radiation) by blackening one set of junctions on which the radiation falls. These junctions become heated, and a voltage is induced that is dependent on the intensity of the incident radiation. It is said that the most sensitive thermopiles are capable of detecting the heat from the flame of a candle several kilometres distant! Since the energy is converted to heat, the only constraint about wavelength is that the 'black' material must absorb most of the radiation at the wavelengths of interest. *Absolute intensity measurements* (whether they are quoted in quantum or photon units on the one hand or in energy units such as joules on the other) ultimately require a measurement of the energy falling on a known area in a given time. The thermopile provides one of the few methods of performing these absolute measurements. One method of calibrating a thermopile for this purpose involves embedding a heating wire of known resistance, R, within the front set of junctions. The voltage, V, produced by the source of radiant energy is first determined, and the source then removed. A measured current, I, is next passed through the heating wire so that exactly the same voltage V is generated by the thermopile as previously. That means that the hot junctions of the thermopile are at exactly the same temperature as they were when heated by the radiant energy. The electrical power required to maintain this temperature is simply I^2R (see the margin of p. 3), measured in joules per second, and this is thus also the

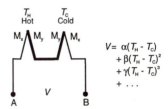

$$V = \alpha(T_H - T_C)$$
$$+ \beta(T_H - T_C)^2$$
$$+ \gamma(T_H - T_C)^3$$
$$+ \dots$$

Fig. 3.17 The thermocouple

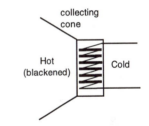

Fig. 3.18 The thermopile arranged for the detection of radiant energy The heavier and lighter lines within the thermopile represent the two different metals

energy that was supplied by the radiation in the same absolute units. Note that this method does away with the need to measure the temperature rise itself, and that it compensates for the heat losses suffered through convection, conduction, and radiation, by the thermopile, because they are identical in both parts of the experiment. One major disadvantage of the thermopile for use in radiation measurements is that it is sensitive to temperature fluctuations from all sources, which makes low-level light detection exceedingly difficult. That is one of the reasons why the photon detectors to be described in the next section are normally used. But the fact remains that absolute calibration of these detectors must have recourse to energy determinations of the kind just explained.

3.6 The detection of light

Vacuum tube photocells and photomultipliers

When radiation falls on the surface of many materials, electrons may be released as a result of the *photoelectric effect*. Sufficient energy must be carried by each quantum of radiation to overcome the *work function* of the particular surface material. As a result, the radiation that is effective in producing photoelectrons has to be of near-infrared wavelengths or shorter (the limit is normally about 900 nm). Photoelectric detectors are, in fact, usually thought of as suitable for the measurement of radiation in the visible and ultraviolet regions of the spectrum. A simple *photocell* for the detection of such radiation combines the photoemitting surface as a cathode with a second electrode, the anode, to collect any emitted electrons, as illustrated in Fig. 3.19. When connected as shown to an external voltage supply, electrons emitted from the cathode will be attracted to the anode, and a current will flow in the external current proportional to the numbers of electrons per second that are collected. The current–voltage relationship is that illustrated in Fig. 3.1, because the current is limited by the supply of charge. For a fixed wavelength of illumination, the limiting current is a linear function of light intensity incident on the cathode.

The quantum efficiency (the fraction of incident quanta, or photons, that yield electrons) depends on the cathode material, and may be a strong function of wavelength, especially near the threshold imposed by the work function of the surface. Materials typically used are pure metals (especially the alkali metals such as sodium and potassium) for the shorter wavelengths, and different kinds of alloy for longer wavelengths. Metal–metal oxide surfaces are also useful, and various kinds of semiconductor material have found increasing application as photocathodes. One point that turns out to be important in choosing a photocathode material is that its work function should be as high as possible compatible with the longest wavelength that is to be measured. *Thermionic emission*, that is thermal production of free electrons from the cathode, can contribute to a *dark current* that is a source of noise in the measurements (see Chapter 5). The production of thermal electrons falls off exponentially as the work function increases. Thermionic emission from

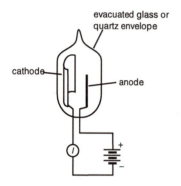

evacuated glass or quartz envelope

cathode

anode

Fig. 3.19 The vacuum photocell

materials that respond only to blue light and shorter wavelengths is normally negligible, while photoelectric devices that respond to the near infrared region often need to be cooled to well below room temperature if the dark current is to be kept at acceptable levels. Another advantage of using a short-wavelength sensitivity cathode is that stray light of longer wavelengths does not interfere with the measurements. For example, a cathode sensitive to short wavelength ultraviolet may be *solar blind*, and used without protection from ambient laboratory light.

Photocurrents from the simple photocell are rather small for low-levels of light intensity, and suitable amplification techniques are needed. It is usual, therefore, to follow the photocathode by an electron multiplier (see pp. 40–41), all in the same evacuated housing, to form a *photomultiplier*. As described earlier, only electrons ejected from the cathode will be amplified (although both photoelectrons and thermal electrons from the cathode will be detected). Photomultipliers can detect very low intensities of radiation, and are used widely in studies of fluorescence and similar phenomena, as well as in the general measurement of light intensities in the laboratory. For the highest sensitivities, electron pulse counting (see p. 41) is used to provide *single photon counting* capabilities for photomultiplier detectors.

Solid-state photodetectors

The insulating properties of a reverse-biassed semiconductor junction diode were discussed in Chapter 1 (see pp. 16–17 and Figs 1.23 and 1.24). If light falls on the junction region, it can generate charge carriers (electrons and 'holes') that will permit a current to flow in an external circuit. This device is then a semiconductor *photodiode*. For applied reverse voltages up to the breakdown value (see Fig. 1.24), the current is virtually independent of voltage (since it is limited by the supply of charge carriers), but is proportional to the incident light intensity. Once again, the limiting-current behaviour of Figs 3.1 and 3.2 is observed. It is, of course, necessary that the light photons possess enough energy to generate the charge carriers. The threshold energy is determined by the semiconductor material and the extent and nature of doping to produce the p- and n-type materials. Many photodiodes are made out of materials that are active for visible and shorter wavelengths, although it is possible to arrange for sensitivity quite far into the infrared region. Just as for red- and infrared-sensitive photocells, thermal generation of charge carriers becomes a problem with these low-energy threshold devices, and it may be necessary to operate them at reduced temperatures. Cooling by liquid helium may even be needed to obtain sufficiently small dark currents for the measurement of low intensities of long-wavelength radiation. Currents from photodiodes are generally quite small for ordinary light intensities, and one useful development has been the fabrication of photodiodes combined with amplifiers in the same integrated circuit package. Because photodiodes can be made very small, it has also proved possible to construct *photodiode arrays* consisting of hundreds or thousands of photodiode elements lined up next to each other on a single small integrated circuit chip. Two-dimensional arrays

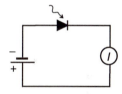

Fig. 3.20 The solid-state photodiode

are also available. Photodiode arrays can be used in a variety of applications in instrumentation, one important use being as a multiple detector element to sit in the image plane of a spectrometer (see p. 78). The response time of most solid-state photodiodes is as good as, or even better than, that of the vacuum tube devices (nanoseconds or less), and they are therefore suitable for use in the investigation of fast transient phenomena. The sensitivity to low light intensities cannot, however, compete with that of the best vacuum photomultipliers in spectral regions where both may be employed.

Cells with semiconductor junctions can also be used to convert light energy into electrical energy. *Photovoltaic cells* are constructed with an insulating barrier layer across which a potential difference appears when light is incident on the cell. Silicon photovoltaic cells illuminated by solar radiation are now widely used as power sources, and drive devices ranging in size from satellite electronics and communications systems to small calculators and watches. Formerly, photovoltaic cells based on selenium were very popular, because their spectral response could be made to match that of the human eye very closely, and the voltage output thus used as a measure of illumination as perceived subjectively. *Photoconductive cells* are also made from semiconductor materials, such as cadmium sulphide. These cells do not have a rectifying junction. Instead, the promotion of valence electrons into the conduction band when sufficiently energetic (short wavelength) light falls on them leads to a decrease in resistance. Cadmium sulphide itself responds to light of wavelength between about 500 and 600 nm, with a quite sharply defined peak at $\gamma \approx 520$ nm. Photoconductive cells show near-ohmic behaviour when connected in a simple circuit (Fig. 3.21). Photovoltaic and photoconductive cells have some advantages over the diode, especially in terms of the simplicity of the external circuitry. They do not, however, in general have either the ultimate sensitivity or speed of response of either the photomultiplier or the solid-state diode.

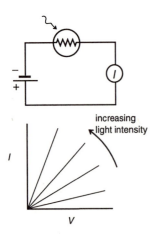

Fig. 3.21 A photoconductive light detector

3.7 Transducers for magnetic fields

Conversion of magnetic fields into electrical signals is required for a variety of purposes in chemical instrumentation, most notably in magnetic resonance spectrometers. One simple transducer is provided by a small coil that is rotated within the field. For a given size of coil and rotation speed, the induced a.c voltage is nearly proportional to the field. A more interesting example with which to conclude this chapter is provided by the solid-state *Hall effect* device. Figure 3.22 shows the principles of construction. A semiconductor material such as indium antimonide or n-type germanium is cut into a thin plate, and an excitation current passed in one direction through it. Pick-up electrodes are connected to the narrow faces at right angles. Normally, of course, the symmetry of the arrangements means that no potential appears across the terminals A and B. If, however, there is a magnetic field perpendicular to the plane of the plate, then a potential difference (the *Hall voltage*) appears across AB, and it is proportional to the strength of the field.

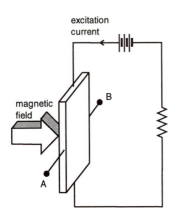

Fig. 3.22 Measurement of magnetic fields using a Hall effect transducer

3.8 Pressure and vacuum measurements

Pressure measurements are important in many chemical experiments. Several methods exist to convert pressures into electrical signals. One of the most obvious is to allow the pressure in the system concerned (liquid or gas) to deform some kind of elastic container and then to measure the deformation with a displacement transducer or strain gauge (Section 3.2). The familiar aneroid barometer produces displacements of the kind described: it consists of an evacuated, flexible capsule, the outer surfaces of which are exposed to the pressure of the system under test (the atmosphere for domestic purposes). The flexible parts of the capsule are usually linked in the domestic device to a series of mechanical levers that operate a pointer. For chemical measurements, it is often convenient to measure differential pressures, so that a simple modification of the barometer allows connection of both inside and outside of the capsule to the two sides of the system under test. One particularly useful way of implementing this technique in electronic instrumentation is to make the flexible part out of a conducting material (a thin metal diaphragm or a metallized plastic membrane), and arrange it to be part of a capacitance displacement transducer as described on p. 37. Figure 3.23 shows the concept schematically. The capacitor has been arranged itself as a differential device, so that, as the pressure changes, the capacitance of one side increases, while that of the other decreases. This arrangement is ideal for connection into an a.c. capacitance bridge, since the bridge becomes unbalanced only if there is a differential change in capacitance. Changes that are common to both sides, perhaps resulting from temperature fluctuations, are cancelled out in the bridge. Suitable signal conditioning can provide an output voltage from the bridge that depends in a known manner on the pressure.

Vacuum techniques and experimentation are extremely important in many aspects of gas-phase chemistry. Once again, it is often necessary to measure the pressure, or at least to establish that it is sufficiently low. One method depends the variation of thermal conductivity of a gas with pressure. The *Pirani gauge* is of this kind. Figure 3.24 shows the principle of construction. A heated platinum wire is contained within a suitable chamber connected to the system under test. The higher the pressure, the greater the heat losses, so that, for a constant heating current, the temperature of the wire will decrease with increasing pressure. In turn, the resistance of the wire can be used as a measure of the temperature (see p. 42). A bridge circuit is normally used to determine the resistance. An interesting small modification to the system just described both improves its performance and makes it adapted to direct readout. The idea is that, instead of rebalancing the bridge when the gas is present by altering the resistances in the opposing arm, the current flowing in the bridge is increased until the bridge rebalances. This rebalancing comes about when the increased current flowing through the test resistance wire exactly balances the heat losses through thermal conduction in the gas, so that the original temperature is restored. Figure 3.25 illustrates the configuration. The current passing through the wire gives a direct measure of the pressure

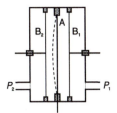

A   = Flexible metal membrane
B_1, B_2 = Fixed capacitor plates

Fig. 3.23 A capacitance manometer based on the displacement transducer of Fig. 3.4. If $P_1 > P_2$, then the diaphragm A flexes towards plate B_2

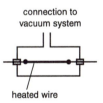

Fig. 3.24 The principle of the Pirani gauge

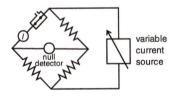

Fig. 3.25 Bridge circuit for the Pirani gauge incorporating modifications for isothermal operation

in the gauge. The method is well adapted to automatic balancing using the techniques to be described in Chapter 4, the output of the null detector being used to control the variable current source to achieve balance. One great virtue of this set-up is that the wire is at a constant temperature, so that the heat losses are always the same. Indeed, the method is used in the *isothermal wire calorimeter* to determine the thermal conductivity of gases. In this application, of course, the gas pressure must be measured in other ways. The Pirani gauge is suitable for use with pressures of about 0.1 to 0.0001 mmHg (remember that 1 atmosphere is 760 mmHg).

At lower pressures again, the *ion gauge* provides a method of determining the pressure. This gauge is a special kind of vacuum tube, illustrated in Fig. 3.26. The spiral wire anode is maintained at a positive potential with respect to a heated filament that emits electrons. The electrons are therefore accelerated, and impact of these electrons with the gas molecules in the tube produces positive ions. These ions are in turn accelerated towards a collector wire which is at negative potential with respect to the anode. The anode and collector are made part of a closed current-measuring circuit, as illustrated in the diagram. For a constant electron beam current (readily stabilized electronically, using the methods outlined in the next chapter), the ion current flowing is directly proportional to the pressure of gas in the gauge. The gauge constitutes a constant current source (current limited by the rate of production of ions: see p. 36), so that the measured current is virtually independent of p.d. between the anode and collector over quite a wide range (cf. Fig. 3.1). The ion gauge is useful for pressures from about 10^{-4} to 10^{-10} mmHg.

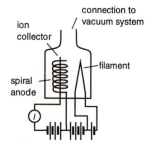

Fig. 3.26 The ion gauge. Electrons are accelerated from the filament to the spiral anode. Collisions with gas molecules generate positive ions that are collected at the central electrode

3.9 Electrochemical measurements

Perhaps the simplest electrochemical measurements are those of electrode potentials that give access to chemical information about condensed-phase ionic systems under equilibrium conditions. A number of thermodynamic and related quantities can be only be obtained readily from e.m.f. measurements of this kind. Conventional electrodes have become supplemented by microelectrodes with tips that are just a few tens of nanometres in diameter. With this kind of electrode, it becomes possible to conduct electrochemical experiments on the interiors of biological cells as well as on other microscopic systems. From the point of view of the instrumentation involved, greater demands than usual are placed on the requirement that no current is drawn (see, for example, pp. 24 and 30–31). Not only must the electrochemical system be maintained under reversible conditions, but the source impedance of a microelectrode may be hundreds of megohms. Feedback amplifiers of the kind described in Chapter 4 provide the solution to this problem.

Other kinds of specialized electrode include those that show a response to specific ions. An obvious example of an ion-specific electrode is the glass electrode used in *pH meters*. The thin wall of glass used in this electrode allows hydrogen ions to diffuse through, but no other species. The nature of this electrode suggests that, once again, the source impedance will be very high, and the appropriate electronic methods must be used to ensure that loading does not invalidate the measurement. Roughly twenty other kinds of ion-selective electrode have been developed.

A different type of electrochemical measurement determines electrode currents that flow in response to given applied voltages across the electrodes. Simple conductivity measurements (see p. 33) are of this kind, but many systems respond in a non-ohmic manner that yields very sensitive analytical information. In general, because currents are being determined, the techniques probe the rates of chemical or electrode processes. A simple example is afforded by the *oxygen electrode*, which provides a useful electrochemical method of determining concentrations of oxygen dissolved in water. Figure 3.27 shows schematically the arrangement used. If the platinum electrode is made about 0.7 volts negative with respect to the reference electrode, oxygen molecules that reach the platinum accept electrons and are reduced. The current flowing in the external circuit is equal to the (diffusion limited) rate of arrival of the oxygen at the electrode, and is thus proportional to the concentration of oxygen in solution. This is a typical current-limited transducer, showing the general characteristics of Figs 3.1 and 3.2. In reality, the constant-current plateau region of Fig. 3.1 is obtained with this electrode over a range of about −0.5 to −0.9 volts.

More sophisticated experiments make use of the measurement of electrode currents as applied voltage is varied. As the voltage is altered, and the potentials at which different electrode reactions occur are reached, quite sharp changes in the current flowing can be detected. The measurements can form the basis of extremely sensitive analytical techniques, and polarography, cyclic voltammetry and anodic stripping voltammetry (ASV) are all of this kind. For example, in ASV a hanging mercury drop is used as a renewable electrode onto which ions are deposited at a relatively large forward potential for a fixed period. The potential is then reversed to increasing values so that each ion is stripped off sequentially. Figure 3.28 illustrates the electrode arrangement used in many of these so-called *potentiostatic* or *voltage-clamp* experiments. In general, a small reference electrode is used to determine the potential of the solution near to the working electrode, and the current supplied through the counter electrode is varied so as to maintain the potential of the reference electrode at the desired value. For varying voltage experiments with this configuration, the (measured) current is altered in order to produce the desired voltage changes. All modern instrumentation of this kind employs electronic control of the current to generate the required voltages. Other electronic circuitry can easily alter the voltage to be attained by a ramp or any other function desired. The control of the current, using suitable feedback between the measured voltage and the current delivered to the counter electrode, is achieved by the methods to be described in the next chapter, and *operational amplifiers* described there (Section 4.4) can also be used to provide the very high input impedances required for accurate electrochemical measurements with high-impedance sources.

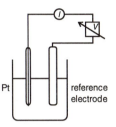

Fig. 3.27 An oxygen electrode

The platinum wire (sealed in glass) forms a microelectrode in the test solution. A counter electrode (e.g. Ag–AgCl) provides a constant-potential reference

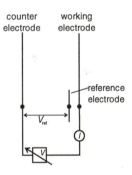

Fig. 3.28. Principle of a potentiostatic electrochemical experiment

The current flowing in the main circuit is varied in order to produce the desired voltage in the solution near the working electrode. The reference electrode monitors this potential

4   Feedback and control

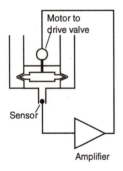

Hot   Cold

To shower

D = Differential expansion element

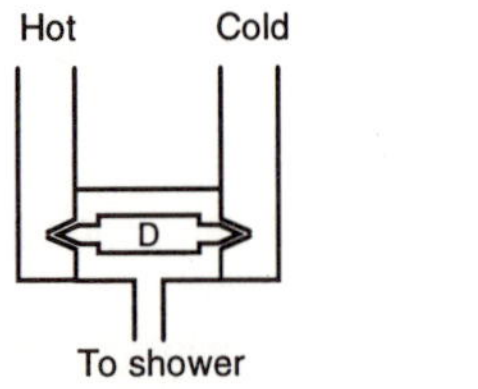

Fig. 4.1 Mechanical thermostat for shower mixer tap

Motor to drive valve

Sensor

Amplifier

The symbol

represents any amplifier, and will be discussed further in Section 4.3. Here it amplifies the error between measured and desired temperature

Fig. 4.2 Electrical analogue of the device shown in Fig. 4.1

4.1 Control systems

Electrical signals can be used to *control* some other property, such as the position of a piece of equipment, the intensity of a light, or the temperature of some system. One way of making this procedure more precise is to measure the quantity being controlled, and use the measurement to make appropriate corrections if the response is not exactly as intended. The return of the measurement to the controlling device is an example of *feedback*. All control systems that incorporate a feedback *loop* can be regarded as types of *servo system*, although the term is often applied specifically to the situation where the position of an object is set electrically, by a motor or other output transducer, and a feedback signal derived from a position-sensitive input transducer (see Section 3.2) to correct the position to that desired. It is in this way that machine tools and other robotic devices can exert forces of many tonnes, and yet move their loads to a hair's breadth of accuracy. In an *open loop* situation, without feedback, the accuracy of the final response relies on the complete transfer function (see pp. 34–35) between initial input and ultimate output. With a *closed loop* incorporating feedback, the accuracy may be determined almost entirely by the calibration of the sensing transducer.

Many control arrangements do not involve electrical signals at all. Examples include simple float and valve devices for keeping liquid levels at the desired value, hydraulic positioning devices (which return information to the operator about the position of the load moved), and a variety of thermostats. It may be instructive to examine a simple non-electrical thermostat so that the principles of operation can be clarified. The shower mixer tap is familiar in everyday life. An 'open loop' mixer requires the operator to judge how much to open the hot and cold taps to give the desired shower temperature. Most readers will know the difficulty in making the correct adjustment, especially if the water supply is subject to vagaries of temperature and flow. A mechanical thermostatic valve eliminates many of these difficulties. Figure 4.1 shows the idea schematically. An element sensitive to temperature is arranged so that if it expands, it tends to block off or reduce the flow of hot water, while if it contracts, it closes off the cold water. In this way, the element remains in the same state of expansion, which is to say that the mixed water must remain at the same temperature. The temperature can be chosen by adjusting the initial position of the element, but, at least within certain limits, fluctuation in temperature or flow rate of the two supplies will not cause the outflow temperature to deviate from that selected.

The electrical analogue (Fig. 4.2) of the thermostatic system is easy to understand. An electrical temperature transducer (see Section 3.5) is placed in the mixed flow in the outlet, the signal is amplified, and it is then used to move the position of the valves in the hot and cold water supplies in the correct sense to maintain the temperature constant. The valves themselves might be actuated by a reversible motor connected to a screw drive. The temperature at which the water flow is maintained is determined by the initial electrical balance point when there is no signal to close either cold or hot supply, and that balance can itself be set electrically to select the temperature.

The thermostatic systems just described embody the elements essential to all closed-loop control systems. These elements are the (i) *measurement* of the quantity of interest (here temperature); (ii) *comparison* of this measurement with the desired value; (iii) the production of an *error response* if the two values are different; and (iv) the use of the error response to take *corrective action*. In electrical control systems, it is usual to amplify the (electrical) error signal so that a small deviation of the measured quantity from its desired value produces a proportionately large corrective driving force. High precision can be achieved in this way, although there may be drawbacks or difficulties as we shall explain in Section 4.6.

Temperature control is, of course, necessary in many scientific experiments, so that the thermostats chosen for illustration have also a very real practical importance. Often, the control of the two separate hot and cold supplies is replaced by a single heater that is driven by the error signal after amplification, with the cooling from temperatures above that chosen being achieved by simple heat losses to the surroundings. If the required temperature is below ambient, a cooling element can be controlled by the error signal, or the apparatus can be cooled to a temperature *below* that desired, and a controlled heater used to bring the temperature up to higher values.

All the devices described so far are *proportional* controllers: that is, the control increases the larger the error signal. Although the word 'proportional' should not be interpreted too literally as meaning 'linearly proportional', this comment applies equally to the mechanical thermostat, hydraulic system, or ball valve as it does to the electrical equivalents. A control strategy that is sometimes simpler just involves turning the driving force on or off according to whether the controlled property is above or below some chosen threshold point. Many temperature controllers are of this type; one example is a room thermostat consisting of a temperature-sensitive switch that may turn a central heating boiler on or off. The great disadvantage of this type of control is that there is almost bound to be some overshoot beyond the desired temperature, because the heating is not cut off until the required temperature has already been reached. We return to this question briefly in Section 4.6.

Several other types of proportional control system are quite easy to devise. A *motor speed controller* might use the chopper disc of Fig. 3.11 (see p. 39) in conjunction with a ratemeter to generate a voltage proportional to the angular velocity. This voltage can then be used to control the power supply to the motor and lock the rotational speed to the desired value. A *light intensity stabilizer* might work in a similar way, with one of the light detectors described in Section 3.6 linked to a lamp power supply. Note carefully that in both cases the power must *decrease* as the transducer output *increases*.

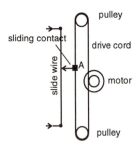

Fig. 4.3 An automated potentiometer: the pen recorder (mechanical outline)

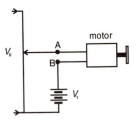

Fig. 4.4 Electrical outline for Fig. 4.3

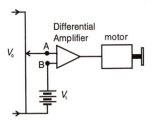

Fig. 4.5 Use of a differential amplifier as an error amplifier

4.2 The potentiometric pen recorder

In Section 2.6, we described the advantages of null methods in making measurements in general, and the virtues of the potentiometer for determining voltages and e.m.f.s in particular. One major disadvantage of the manual potentiometer is the time taken to balance it; the speed and accuracy depend on the skill of the operator. An automated version of the potentiometer, on the other hand, allows electrical signals to achieve the balance point with both high speed and high accuracy. The simple electrical circuit of the potentiometer is given in Fig. 2.20 (p. 31). A way of implementing this arrangement automatically is shown in outline in Fig. 4.3. An electric motor is used to move the sliding contact up and down the wire via a drive cord and set of pulleys. The next step is to connect up an electrical circuit such as that indicated in Fig. 4.4. Points A and B are those that were connected to the null-indicating meter in Fig. 2.20. They are now arranged to drive the motor to bring the slider down the slide wire if the voltage at A is bigger than that at B, and up the wire if it is smaller. The motor comes to rest when the potentials at A and B are identical, and the position of the slider on the wire is exactly that required to obtain balance. One could imagine such an arrangement in which a small d.c. motor (whose direction of motion depend on the polarity of voltage applied) driving a very low mass and low friction slider. In reality, the difference in voltage between A and B is always amplified in an *error amplifier*, as shown in Fig. 4.5 A suitable amplifier is obviously a *differential amplifier*, some kinds of which are discussed in Section 4.4. Its inclusion does not alter the principle of operation, but it allows the motor to exert a large torque for small displacements of the slider from the balance point, and, in addition, it can prevent current being drawn from the test voltage source while the potentiometer is balancing itself.

It is just one simple step from this automated potentiometer to the construction of a pen recorder, and that is to attach a suitable pen to the mechanical link to the sliding contact. The pen thus moves in exact relation to the slider; the differential amplifier provides the power necessary to drag the pen along the recording paper. A *strip chart recorder* employs a constant speed motor to drive the recording paper under the pen, and thus, in its simplest form, provides a record of voltage at point B as a function of time. An *X–Y recorder* is set up with two sets of potentiometers with their sliders at right angles, and arranged so that one potentiometer is carried on the other. The pen writes on a flat-bed paper sheet to plot *Y* voltage as a function of *X*.

4.3 Feedback

Negative and positive feedback

At the end of Section 4.1, it was emphasized that the transducer voltage and the phenomenon being controlled must be linked in such a way that an increase in the transducer output will tend to decrease the stimulus producing it. Thus an increase in temperature must be arranged to decrease the heating process in a thermostat, an increase in angular velocity must reduce the power supplied to the motor in the speed controller, and an increase in light intensity must reduce the power applied to the lamp in the intensity stabilizer. Feedback applied in this way is *negative feedback*. Negative feedback in a control or servo system tends to stabilize it, because any change is always opposed by the sense of the control voltage or current. *Positive feedback* is also occasionally used in some applications, and it has the opposite effect, that of *de*stabilizing the system, usually with the result that it locks to one extreme or the other of its range of response, or else goes into oscillation. We shall meet some examples in Section 4.5. In all the negative feedback cases, with which most of this chapter is concerned, the concept is always that of error correction: the actual response is compared with the desired response and any deviations are employed to reduce the discrepancy.

Feedback within the electronic system

So far, the discussion of this chapter has concentrated on closed feedback loops in which the loop includes both electrical or electronic devices (input transducers and amplifiers), and an output transducer producing a non-electrical response (heat, motion, light, and so on). Negative-feedback control of the purely electronic or electrical functions also conveys very substantial advantages, including the expected one of improved stability. Negative feedback operating solely within the electrical system can still be regarded as providing servo control, but now of the electrical output responses rather than of the other physical responses that we have considered.

A convenient point at which to start this short survey is to consider the basic elements of a *voltage stabilizer*. Many laboratory experiments require a source of voltage that does not change either when the external power supply fluctuates, or when the load imposed by the experiment alters. The circuit of Fig. 4.6 shows how an electronic servo regulator might work. A differential amplifier, such as that introduced in connection with Fig. 4.5, and described in more detail in Section 4.4, amplifies the *difference* in voltages between points A and B. Input B is connected to a source of constant voltage (perhaps a standard cell, but more probably a source based on the zener diode described on p. 17). Input A is connected to, or samples, the output voltage of the complete regulator at point O. The unregulated input supply is passed through a voltage controller to the final output. This voltage controller is essentially an electronically variable resistor; it will be made up of transistors capable of carrying the required currents, but its circuitry is immaterial in the present context. The essential feature is that the effective resistance increases as the control voltage applied at the point C increases. Now the point CO is also that at which the amplified error output from the differential amplifier appears. If there were a tendency for the final output at O to increase, the voltage at C would decrease, thus preventing the change at O. The reverse sequence prevents decreases in voltage at O, which is thus stabilized to a value linked to the reference voltage, V_{ref}. The stabilization is effective regardless of whether the tendency to output change results from changes in

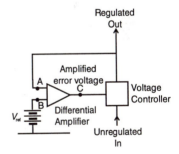

Fig. 4.6 Electronic servo-regulator voltage stabilizer

the input voltage or alterations in the load. The servo system operates to compare the output voltage with V_{ref} and take corrective action if the relation between output and reference is not that intended. Suitably large currents can be carried by the controller, while the reference supply is not called on to supply substantial currents, and so can be chosen or designed for stability rather than current-carrying capacity. Common use is made of *integrated-circuit regulators* in which the reference unit, amplifier and controller are fabricated on one silicon wafer.

Negative feedback is very frequently used across the amplifying elements themselves. Figure 4.7 indicates the essentials of a voltage-feedback circuit. An amplifier is assumed to have a voltage gain of A: that is, a voltage V_i applied to the input terminal produces an output voltage of $V_o = AV_i$. In the closed-loop circuit, a fraction β of the output voltage, V_o, is fed back and *subtracted* from the input voltage, as shown in the figure. Thus, if the input voltage to the complete amplifier is V_i, the voltage applied to the point X is $(V_i - \beta V_o)$, and it is this voltage that is multiplied by the gain A to produce the output, V_o. As shown beneath the figure, the immediate conclusion is that the gain of the complete system, A' say, is equal to $A/(1 + A\beta)$.

The point of all this is that A' is a less than linear function of A. Simple amplifiers usually exhibit open-loop gains (that is, the gains *without* feedback) that are dependent on a variety of factors, including the particular transistors used, the operating voltages, and the temperature. The gain with feedback varies less than the gain without it. Indeed, if A is sufficiently large that $A\beta \gg 1$, then $A' = 1/\beta$. That is, the overall voltage gain of the system is dependent only on the fraction of output voltage fed back to the input, and *independent of the open-loop gain of the amplifier*. The feedback fraction can be determined with simple 'passive' components such as resistors, so that the gain can be selected reliably and reproducibly with these components alone, and extraneous factors such as supply voltage fluctuations become of minor consequence.

There are several other advantages to be obtained by the use of negative feedback. One familiar example is in amplifiers intended for high-fidelity reproduction of sound, in which feedback is used to reduce *distortion*. The same principle applies to reduction of distortion in many examples of laboratory instrumentation. Such distortion might mean that a pure sine wave applied to the input is no longer a pure sine wave when it reaches the output of the amplifier, perhaps because of *non-linearity* in the amplifier transfer function. Much of the distortion is likely to be introduced in the output stages of an amplifier, especially if relatively large powers are required, because it is in these stages that there are large swings of voltage and high currents are carried. If negative feedback is applied to the amplifier, and the gain of the first (low signal) stages of the amplifier increased to compensate, then the output swings will be as large as before. The distorted component of the output would now constitute an error signal at the input in a sense such as to counteract the distortion. In other words, the feedback introduces a voltage into the output of opposite sign to the distortion which is thus almost exactly

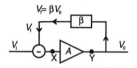

Fig. 4.7 A voltage-feedback amplifier.

Voltage at X = $V_i - V_f$
         = $V_i - \beta V_o$
Voltage at Y =
     $A \times$ voltage at X
         = $A(V_i - \beta V_o)$
But this voltage is also = V_o, so that
$A(V_i - \beta V_o) = V_o$, or
     $V_o = V_i A/(1 + A\beta)$
The overall gain of the circuit, A' (= V_o/V_i) is thus given by
     $A' = A/(1 + A\beta)$
If $A\beta \gg 1$, $A' = V_o/V_i \approx 1/\beta$

cancelled. Another change in circuit behaviour that follows application of negative feedback and that is nearly always advantageous concerns the impedances. The *input impedance increases* and the *output impedance decreases*. As a consequence, amplifiers with feedback both load the input circuit less, and are less affected by the load attached to their outputs, than the corresponding amplifiers without feedback. The derivations of the impedances with feedback will not be given here, but they are simple, and the reasons can be seen qualitatively. At the input end, any voltage applied is partially balanced by the feedback voltage of opposite sign, so that less current flows than without the feedback. At the output, an increased load (that is, decreased load resistance) would normally reduce the output voltage. However, with feedback in the circuit, this reduced output would also mean a smaller feedback, so that the drop in output voltage will be less than without feedback.

4.4 The operational amplifier

Differential amplifiers were introduced briefly in connection with Figs 4.5 and 4.6. Used in conjunction with particular feedback configurations, amplifiers of this kind form the basis of *operational amplifiers*. The name derives from the ability of such amplifiers to perform set mathematical operations, such as addition, subtraction, integration, and differentiation, a capability that was of particular use when analogue computers had not been virtually superseded by digital ones. However, integrated-circuit operational amplifiers have become the basic building blocks of much analogue electronics, and also form the interfaces with the digital world as well, and some of their properties and uses will be explored in this section.

A requirement for operational use of an amplifier is that it responds from zero frequency (that is, d.c.) upwards. When valves or discrete transistors had to be used, construction of d.c. amplifiers offered many difficulties, and *drift* (slow change of d.c. output level with time) as a result of temperature and supply voltage fluctuations, and even because of component ageing, was a particular problem. Development of integrated-circuit technology has revolutionized the construction of d.c. amplifiers, because the ability to make all the components together on a single semiconductor 'chip' means that it is much easier to build an electrically balanced and optimized circuit in which almost all of the potential drifts are made to cancel out. Modern integrated circuits for use as operational amplifiers are not only small, but are often very cheap, yet they may contain the equivalent of from 15 to many more individual transistors. The desirable (and attainable) qualities, apart from freedom from drift, include high open-loop gain (see p. 54) and a high input impedance. The use of a differential amplifier arises in part because of the need to subtract feedback voltages from the input values as indicated in Fig. 4.7. With a differential amplifier, the subtraction process is an inherent property of the amplifier. Figure 4.8 shows the usual symbol for a differential amplifier, in which '+' and '−' signs have been placed at the inputs. The '+' input is the *non-inverting input*, while the '−' input is the *inverting input*.

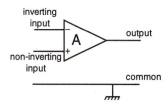

Fig. 4.8 Some terms associated with the differential amplifier

Within the output voltage capabilities of the amplifier, the amplifier should produce an output voltage, V_o, of AV_i for V_i applied to the non-inverting input, and of $-AV_i$ when the same voltage is applied to the inverting input, where A is the open-loop gain of the amplifier. In general terms, $V_o = A(V_+ - V_-)$ if V_+ and V_- are the voltages at the non-inverting and inverting inputs: the output is proportional to the *difference* in input voltages. (Note that the positive and negative symbols do not indicate positive or negative voltages, but only the amplifier input to which they apply). Implicit in these remarks is the idea that if $V_+ = V_-$ the output voltage is zero, so that if the two input terminals are connected together, no output voltage should appear regardless of the voltage that might be applied to the joined inputs. This property of rejecting common input voltages is *common mode rejection*; a numerical measure of the rejection is expressed as a *ratio* (and abbreviated CMRR). A high CMRR is another highly desirable property in a differential amplifier intended for use as operational amplifiers. One other feature of Fig. 4.8 needs to be explained here. All the voltages discussed and measured have to be referred to some point. It is usual for the reference to be the circuit *common*. This common line may then be considered to be at zero volts for the purposes of measurement. It may or may not be connected to physical *ground* or *earth* potential (a potential nominally the same as that of the local ground in the geographical sense).

Integrated-circuit amplifiers can easily possess open-loop voltage gains of 10^5 or more. The maximum output voltage cannot be more than the voltage of the power supply energizing the amplifier, which is typically ±15 volts for the types of amplifier under discussion. The amplifier is thus likely to *saturate* or *limit* under open-loop conditions with differential inputs exceeding a value of the order of tenths of a millivolt. Use is sometimes made of the limiting behaviour (see Section 4.5), but, in most applications, the feedback is arranged to keep the amplifier response linear.

A great variety of circuits can be constructed from the simple operational amplifier; we can look only at a few of the most important types of configuration. We shall make the approximation that the amplifier behaves ideally, and that it has an input impedance and a gain each approaching infinity. The feedback provided reduces the closed-loop gain to the value required, and ensures that the amplifier output is within the linear regime. It follows that, in all the examples, the voltages at inverting and non-inverting inputs are *almost* the same, and interpretation of circuit behaviour is much simplified by assuming that the voltages are identical. The easiest circuit with which to start is the *voltage follower*, shown in Fig. 4.9. In this circuit, *all* the output voltage is fed back to the inverting input: that is, the parameter β introduced on p. 54 is unity. Since, as we have just said, the voltages at the two inputs are virtually identical, the output voltage, V_o, must adjust itself to be equal to the input voltage, V_i, if the output is connected to the inverting input. The voltage gain is thus unity. This qualitative analysis is confirmed by the simple formula $V_o/V_i = 1/\beta$ (see p. 54 and the legend to Fig. 4.7) for $\beta = 1$. The output must *follow* whatever voltage is applied to the inverting

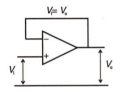

$V_i = V_o$

V_i

V_o

Fig. 4.9 The voltage follower

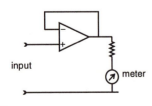

input

meter

Fig. 4.10 A high-impedance voltmeter

input of the amplifier, and it is this behaviour that gives the circuit its name. An amplifier that does not alter the input voltage might be thought to be of rather limited use, but nothing could be further from the truth. The value of the amplifier is that it acts as an *impedance converter* with the very precise voltage transfer characteristic just described. Consider now the load placed across the output. The output has to stay at the input voltage regardless of the load, so the amplifier just has to supply more current to the load. In this idealized view, the output impedance is zero; a real amplifier cannot supply infinite currents of course, nor does it have infinite gain, but the principle remains the same. At the input end of the amplifier, no current can flow between the non-inverting and the inverting input because they, too, are at the same potential. Thus the input impedance is infinite; finite, but very small, currents may flow in a real amplifier. The ability to supply currents to a low-impedance load while presenting a negligible load to a source is valuable in a variety of applications in instrumentation. Figure 4.10 shows how a voltage follower might be coupled to the simple meter circuit of Fig. 2.7 to produce a high input impedance voltmeter. Such a meter could be used to make e.m.f. measurements on electrochemical cells, for example, without appreciably perturbing the system. The voltage follower is thus a *buffer* of the kind whose use was suggested on p. 23. The same circuit is also used, either as a separate circuit element, or as part of a larger integrated circuit, as the buffer preceding the ADC part of a digital voltmeter (see pp. 24–25).

If voltage gain is required, then the trick is to feed back only part of the output voltage, thus making $\beta < 1$, and $V_o/V_i = 1/\beta > 1$. Figure 4.11 shows the circuit for this kind of *non-inverting voltage amplifier*. A potential divider takes off a fraction $\beta = R_1/(R_1 + R_2)$ of the output voltage, which is thus the inverse of the gain of the unit. The same result comes, of course, without the use of the gain formula, but by noting that the voltage at the inverting input is $V_o \times R_1/(R_1 + R_2)$ and that this must therefore also be the voltage at the non-inverting input. The gain of the amplifier with feedback is determined by the resistors alone, and can thus be set precisely. The exact gain of the differential amplifier itself is immaterial so long as it is high enough for the approximations made (see legend to Fig. 4.7) to apply. As in the case of the voltage follower, the input impedance of this amplifier is very high and its output impedance low. Because it is non-inverting, the output voltage is of the same polarity, and changes in the same sense, as the input voltage; for a.c. signals, the output voltage is *in phase* with the input signal.

Inverting amplifiers are employed, for example, if the power supplied to some experiment must decrease as some input signal increases, or if it is necessary to arrange for a phase shift of 180° in an a.c. signal. The process of inversion is, of course, ultimately at the heart of all negative feedback and servo control devices, and it is thus of very great importance. Figure 4.12 shows the basic operational amplifier circuit. Both input and feedback are connected via resistors to the inverting input. As we shall show shortly, an algebraic summation (that is, a summation with due consideration of sign) of currents occurs at this input, and the point marked S is thus the *summing*

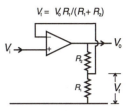

$V_i = V_o R_1/(R_1 + R_2)$

Fig. 4.11 A non-inverting voltage amplifier

$$V_o = V_i(R_1+R_2)/R_1$$
$$V_f = V_o \times R_1/(R_1+R_2)$$
but $V_i = V_f$, so that
$$V_i = V_o \times R_1/(R_1+R_2), \text{ and}$$
$$A' = V_o/V_i$$
$$= (R_1+R_2)/R_1$$

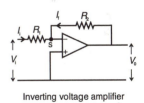

Inverting voltage amplifier

Fig. 4.12 An inverting voltage amplifier

In order that the summing junction S is maintained at the common potential, I_f must be equal to $-I_i$. Thus,
$$I_f = V_o/R_2$$
$$I_i = V_i/R_1, \text{ and}$$
$$(V_o/R_2) = -(V_i/R_1), \text{ or}$$
$$A' = V_o/V_i = -(R_2/R_1)$$

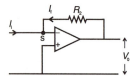

Fig. 4.13 Current to voltage converter

$$V_o = I_fR_2$$
$$I_f = -I_i, \text{ so that}$$
$$V_o = -I_iR_2$$

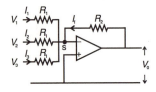

Fig. 4.14 The analogue adder

$$I_1 + I_2 + i_3 = -I_f$$
so that
$$V_1/R_1 + V_2/R_1 + V_3/R_1 = -V_o/R_2$$
Thus
$$V_o = -(R_2/R_1) \times (V_1 + V_2 + V_3)$$

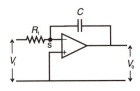

Fig. 4.15 The integrator.

The total charge delivered in time t is

$$\int_0^t I_i(t)t = \frac{1}{R_i}\int_0^t V_i dt$$

This charge must be balanced by the charge on the capacitor ($q = CV_o$: see Fig. 1.13) Thus

$$V_0 = -\frac{1}{R_iC}\int_0^t V_i dt$$

junction of the circuit. The non-inverting input is connected to the circuit common (or ground, or earth); as stressed earlier, the inverting input and thus S must be at almost the same potential, and S is said to be at *virtual common* (or *virtual ground*, or *virtual earth*) potential. The circuit behaviour can be analysed as indicated in the legend by considering the currents flowing into the summing junction. Since the amplifier itself draws no current, the flow of current into the junction from the source must be exactly equal to the flow of current into it from the output through the feedback resistor, *but opposite in sign* so that the currents are cancelled. The overall gain of the circuit, V_o/V_i, is thus equal (see figure legend) to the ratio of the resistances $-R_2/R_1$, the negative sign reflecting the inverting nature of the amplification. The output impedance is again very low, because the output has to adjust itself to keep the feedback sufficient to balance the input. The input impedance does not approach infinity in the way that it does with the non-inverting circuits. Rather, it would be exactly equal to R_1 with an ideal differential amplifier, because S is always kept at zero potential with respect to the circuit common. Indeed, because it is the current flow into S that determines the output voltage, the circuit can be used as a *current amplifier* or a *current to voltage convertor*, as indicated in Fig. 4.13. Such an amplifier produces an inverted output voltage numerically equal to the input current multiplied by the feedback resistance. Use of a high feedback resistance thus allows the measurement of correspondingly small currents, and the configuration is extremely valuable in processing the output of the current-source transducers (see pp. 35–36), such as the charged particle detectors and photocells described in Sections 3.4 and 3.6.

It is easy to see now how the operational amplifier can be used as an *analogue adder*. Figure 4.14 shows the circuit of Fig. 4.12 modified to have three separate input voltages. The three input currents must therefore balance the feedback current to keep the summing junction at zero potential. The input resistors have been chosen for simplicity in this case to be equal; the algebra shows that the output voltage is a function of the *sum* of the three input voltages, and the device has thus added them. Indeed, if R_1 is made the same as to R_2, the output is directly equal to the sum of voltages. Although analogue computers are not in fashion, there are still many times when it is desirable to display the sum of several outputs in this way, or to add three signals precisely for further processing. Another mathematical operation that is easily performed is that of *integration*. A capacitor (see pp. 11–12) is used in the circuit (Fig. 4.15) to store charge; the output voltage is related to the accumulated charge (that is, its integral). The principle indicated is used in a variety of applications not so much for mathematical integration, but to alter the frequency response and introduce phase shifts into a.c. amplifiers. Similar remarks apply to the *differentiator* (a circuit in which the capacitor and resistor of Fig. 4.15 are interchanged in position). The facility with which frequency responses can be chosen with these types of circuit has encouraged the use of operational amplifiers in a variety of *active filters* (compare with pp. 14–15) in which the response can be tailored to the needs of the application.

4.5 Positive feedback

There are some circuits in which positive feedback is employed. While negative feedback tends to stabilize the system, positive feedback *destabilizes* it so that a small displacement from an equilibrium situation is exaggerated until the system moves as far from equilibrium as possible. Consider the circuit of Fig. 4.16. We have seen that with a high-gain amplifier operated in an open loop, a differential input of less than a millivolt can produce an output swing up to the saturation limit of the amplifier. With positive feedback added, this behaviour is reinforced and modified. A small signal applied to the inverting input is amplified to give a large output of opposite sign. Since all of this output is fed back to the non-inverting input, the differential input voltage is greatly increased, and a yet larger output is produced. The output very rapidly reaches its saturation value, with a polarity opposite to that which started the sequence. A small negative voltage produces a limiting positive output, and vice versa. A short duration pulse of appropriate polarity can thus be used to set one of the two stable limiting output states, so that this *bistable* circuit acts as a memory *latch*. Such devices are at the heart of digital electronics, and can be extended from memory elements to simple counters and dividers. In the simple circuit, it may not be possible ever to reverse the changes without disconnecting the circuit. A small modification is shown in Fig. 4.17: not all the signal is fed back, but only a fraction $\beta = R_1/(R_1 + R_2)$: the circuit is the positive feedback analogue of the negative feedback circuit of Fig. 4.11. The fraction β of the saturated output voltage, V_s, appears at the non-inverting input. Suppose the output voltage is initially positive. An voltage *more positive* than βV_s applied to the inverting input will cause the output to change state (go negative); in this latter state, an input voltage *more negative* than βV_s (itself now negative) will reverse the output state again. There is thus a gap, or *hysteresis*, between the input voltages that effect the state changes. Devices based on this circuit (the *Schmitt trigger*) are extremely useful as *threshold comparators* or *discriminators*. For example, it may be required in an experiment to distinguish between output levels that are 'high' or 'low'. With the circuit of Fig. 4.17, it is possible to make sure that small random fluctuations in output (noise: see Chapter 5) do not falsely trigger the transition. Rather, the hysteresis can be arranged so that the 'high' is recorded only for signals that are larger than 'low plus noise', and that low is recorded only for signals lower than 'high minus noise'. Connecting R_1 to a reference voltage source rather than to common allows the voltage at which the discriminator acts to be set according to the needs of the experiment.

Yet other uses of positive feedback devices include a variety of *oscillators*. Figure 4.18 shows one possibility. A frequency-dependent phase-shift network connects the output to the inverting input. If an a.c. signal is shifted by 180°, a negative signal will appear at the inverting input as the output goes positive, and vice versa. Inversion in the amplifier reinforces the output, so that there is net positive feedback. A sustained increase in amplitude can thus build up at that frequency for which the phase shift is exactly 180° and oscillations occur.

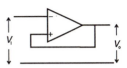

Fig. 4.16 An amplifier with positive feedback

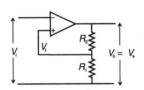

Fig. 4.17 The Schmitt trigger threshold sensing circuit

Assuming that $A\beta \gg 1$ as hitherto,
$$V_f = V_s R_1/(R_1+R_2) = \beta V_s.$$
A value of V_i greater than the positive magnitude or smaller than the negative magnitude of V_f will result in a change of state. There is thus a hysteresis zone of $2\beta V_s$ over which V_i can change without triggering a change of state

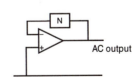

Fig. 4.18 An oscillator circuit. N is a frequency-dependent network

4.6 The stability problem

Oscillations may not be deliberate, but come about accidentally because there are unexpected in-phase feedbacks with sufficient gain to sustain the oscillation in what is meant to be a non-oscillating circuit. The effect can render an otherwise stable circuit useless. Faults in design, construction or components can lead to instabilities, and it is essential that they are eliminated in any instrumentation that is to work reliably. The cures include examination of loop gains and phases at all appropriate frequencies, and the elimination of stray impedances and feedback paths. The problem is compounded where there are non-electrical components such as heaters, motors, or lamps in the feedback loop. In these cases, the non-electrical response may be sufficiently slow that it introduces a delay equivalent to 180° at some relatively low frequency. The entire system may then oscillate uncontrollably.

A related issue in the servo control applications is that the electrical driving force must be sufficient to push the non-electrical system to the desired end result. With a simple proportional controller, the drive becomes increasingly small as the error signal becomes smaller and the end condition might only be reached after inordinately long times. The time taken to reach the 'equilibrium' condition can be shortened by increasing the loop gain, so that the electrical drive is high for a small error signal, but that increases the likelihood of oscillation. In real devices, also, the non-electrical components show a marked inertia. This inertia might be thermal in the case of a heater or mechanical in the case of a motor driving the potentiometer of the pen-recorder. In general, provision of sufficient loop gain to bring the system rapidly towards the correct condition will cause there to be an *overshoot*. This overshoot is then corrected by the servo system, and a (smaller) overshoot occurs in the opposite direction. The process repeats itself, with decaying amplitude, so that the changes correspond to *damped oscillations*, illustrated for a motor-driven pen recorder in Fig. 4.19 as the response (b) to an abrupt change in voltage (a). Sustained oscillation is an extreme form of this behaviour, sometimes referred to as the servo system *hunting* for its correct condition (response (c)). Time constants (and phase relationships) in the electrical and non-electrical parts of the system must be correctly matched, so that high gain (and thus precision and speed) can be combined with efficient *damping* that prevents the occurrence of oscillation or significant overshoot. The damping modifies the *control law* from that of true proportionality, because the gain is different at different frequencies. One particularly useful control law uses *proportional, integral, and derivative (PID)* components. The integral part provides high gain at low frequencies to maintain the long-term accuracy; the derivative part depends on the rate of change of error signal, allowing a large control power to compensate for rapid changes. Many commercial controllers use microprocessors to provide a programmed control law. The causes of, and cures for, instability are complex, but some understanding of them can help users of instrumentation to turn a black art of adjustment into a process that has a rational basis and a chance of success.

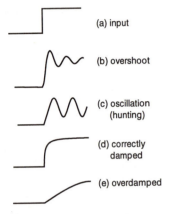

(a) input

(b) overshoot

(c) oscillation (hunting)

(d) correctly damped

(e) overdamped

Fig. 4.19 Responses in a pen recorder

5 Noise and its reduction

5.1 Noise and experiments

By analogy with sound, where *noise* implies the presence of unwanted sounds, the term as applied to electrical systems also means that there are voltages or currents present that are not related to the quantity being investigated. Just as with sound, the noise detracts from the information that can be gathered from material to which the hearer is listening. The reduction of information means a loss of precision that can be regained only by conveying the message more slowly or by repeating it as many times as necessary (which also means a slower process). In Chapter 1, it was pointed out that noise always accompanies the signal in any real piece of instrumentation (see pp. 9–11). Even if there are no 'avoidable' sources of noise (see later), there remain inherent sources of noise that thus serve to limit the precision of measurements. The *signal-to-noise ratio* (*S/N*) is a measure of the relative magnitudes of the wanted signal and the unwanted noise. The smaller the *S/N* ratio, the more difficult it is to make the required measurement. The only way to restore the quality of the information is to spend longer on making the measurement.

In many applications of electronic instrumentation in science, the signals are often very weak and the noise relatively large. The problems of *S/N* ratios can thus be very important, and proper handling of the available material may sometimes mean the difference between success and failure of an experiment. A skilled experimentalist will attempt both to reduce the noise to as near to the theoretical limit as possible, and to ensure that signals that are available are handled and processed in such a way as to ensure that the information in them is degraded or wasted as little as possible. This chapter is intended first to give an outline of what the sources of noise are and the extent to which the noise may be reduced, and secondly to illustrate methods for making the best of the initial *S/N* ratio and of enhancing the precision of the measurements. An understanding of these principles can often guide the user to a more effective employment of instrumentation in tackling a problem.

Sources of noise

Before proceeding further, it is essential to investigate the different sources of noise. For our purposes, it is convenient to distinguish between three distinct kinds of noise, which are

(i)   *fundamental noise*;
(ii) *excess noise*; and
(iii) *interference*.

Fundamental noise arises from the physical properties of electrical systems, and cannot be eliminated. Excess noise is a term applied to noise generated from within the instrumentation, possibly as a consequence of imperfections in the components of the apparatus. Improvements in design can bring a reduction in noise of this type. Finally, interference arises from outside the instrumentation itself; in principle, it can be eliminated entirely, although the practice often proves difficult and depends on experience and intuition. The characteristics, and especially the frequency spectrum and time distribution, of the different sorts of noise aid in discriminating between signal and noise (Section 5.3), and thus in improving *S/N* ratios (Sections 5.4 to 5.6).

Fundamental noise

Formula 5.1 The mean square of the thermal noise voltage is given by the equation

$$\overline{V_n^2} = 4kTR\Delta f$$

where k is the Boltzmann constant, R the resistance in which the noise is being generated, and Δf is the bandwidth in which the noise is observed.

Motions of discrete charge carriers, such as electrons, in electrical circuits give rise to the fundamental noise. There are two important types. The first arises directly from the thermal motions of electrons in conductors, which generate a noise voltage even in the absence of a signal current. This noise is *thermal noise*, and is also known as *Johnson noise* or k*T noise*. The second type of fundamental noise is *shot noise*, and it arises from the statistical nature of the flow of a signal current. The name is said to derive from the sound made when a large quantity of lead shot is poured. Shot noise shows up as random fluctuations superposed on a mean level of the signal current.

Thermal noise occurs in resistors and similar devices that are at temperatures above absolute zero, and result from the thermal agitation (Brownian motion) of all particles. Fluctuations in the velocities of the charge carriers gives rise to a fluctuating unbalance of charge across the resistor and thus to a noise voltage. Application of the laws of thermodynamics (and especially the equipartition theorem) allows an average value for the noise energy or power to be calculated. The electrical equivalent is a mean square noise voltage, $\overline{V_n^2}$, for which the equation is given in Formula 5.1. Substitution of numbers into the equation shows that the square root of $\overline{V_n^2}$ (that is, the rms noise voltage: see p. 24) should be about 1.3 µV in a 10 kHz bandwidth for a 10 kΩ resistor at room temperature (300 K). The quantity $(\overline{V_n^2}/R\Delta f)$ is the noise power per unit frequency interval, or the *noise power density*. One important conclusion to be drawn from the equation is that the this power density is *independent of frequency* (see Fig. 5.1): the noise is said to be 'white', this time drawing on an optical analogy. In fact, the only quantities remaining in the expression for the power density are the Boltzmann constant, k, and the temperature, *T*. It should be noted, however, that the derivation is based on *classical* thermodynamics, and when the equipartition theorem becomes inapplicable (at energies corresponding to frequencies of more than a few hundred GHz) the noise ceases to be white.

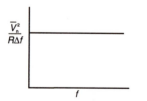

Fig. 5.1 The thermal noise power density spectrum. The ordinate, $\overline{V_n^2}/R\Delta f$, is the power density (i.e. the power per unit frequency interval)

Thermal noise can be minimized at source in two ways. The first is to arrange for resistances in the critical (low signal) parts of the circuits to be kept as low as possible. Further improvement may be achieved by operating these critical parts of the circuit at low temperature; cooling of ultra-sensitive amplifiers by liquid helium is not unknown.

Shot noise is a direct consequence of the way in which current is carried by individual charges, negative or positive, that ultimately depend on the presence or absence of electrons. Even quite small currents involve very large numbers of electrons (one microamp, for example, corresponds to a flow of more than 6×10^{18} electronic charges per second), and the effects of the 'granularity' of the current are not normally apparent. However, when small currents are made up of independent charge carriers, as they are, for example in the currents of charged particle detectors (Section 3.4) and many photon detectors (Section 3.6), and at semiconductor junctions, the current pulses are essentially random. Together, the arrival of the charges constitutes a mean current, but the randomness means that there is a fluctuation about the mean. Formula 5.2, which shows the mean square current noise, can be derived from simple statistics. The noise is again white, showing the frequency-independent characteristics similar to those of Fig. 5.1. Note that, although the *absolute* shot noise current increases with the signal current, I, the *S/N* ratio decreases inversely as the square root of I. For a value of $I = 1$ μA and $\Delta f = 1$ kHz, the shot-limited *S/N* ratio is about 56 000, easily enough to satisfy most requirements in chemical instrumentation. Currents a million times smaller are not uncommon, however, and the *S/N* ratio will then drop to 56, adequate for many purposes, but corresponding to a limiting precision of the measurement of only just better than two percent.

Dark currents in photodetectors (see pp. 44–46), and thermionic leakage currents in other kinds of semiconductor, can be thought of as related to both resistive thermal noise and shot noise. The currents have their origins in the thermal generation of charge carriers, the numbers depending in a way described by the Boltzmann distribution on the temperature and the work function (or band gap) of the material concerned. If the dark current were absolutely steady, it might merely provide an offset to the current to be measured. Unfortunately, though, the dark current is itself subject to statistical shot noise which is superimposed on the shot noise associated with the signal itself. The practical remedies are to use cooled devices (and the avoidance of these thermionic currents may be a better reason for cooling rather than just the reduction of Johnson noise), and to select cathode or junction materials that have the highest possible work function or band gap consistent with the application envisaged. For example, to detect ultraviolet radiation, a photoelectric device with a cathode that has a long-wavelength cut-off in the blue region of the spectrum is more suitable than one whose sensitivity extends into the deep red.

Excess noise

Noise beyond the inevitable thermal and shot components is regarded as excess noise. The changes in leakage currents with changing temperature just discussed should therefore probably be counted as excess noise. Temperature sensitivity of individual electronic components, circuit elements, or even complete instrumentation certainly contributes to excess noise. In the context of complete systems, for example, an instrument such as a spectrophotometer

Formula 5.2 The mean square shot noise current is given by the expression

$$\overline{I_n^2} = 2eI\Delta f$$

where e is the charge on the electron and I is the (mean) signal current. The signal-to-noise ratio (*S/N*) is thus given by the equation

$$S/N = (I/\overline{I_n^2})^{1/2} = (I/2e\Delta f)^{1/2}$$

might show sensitivity to geometric distortions as the temperature alters, and changing temperature might also affect the intensity of light sources as well as the dark current of detectors and amplifiers. Contacts made between different metals can be a source of thermoelectric e.m.f.s (see p. 43) if there are temperature differentials present. Poor contacts and bad soldering can lead to fluctuating circuit resistances. These kinds of problem can be avoided, or at least alleviated, by proper circuit design and constructional technique, and, if necessary, by thermostatic control of the temperature. Excess noise introduced by the circuit components themselves can be more difficult to deal with. Carbon composition resistors are often noisy, probably because of their granular construction. Semiconductor devices, including diodes, transistors, and integrated circuits, often introduce their own excess noise, possibly because of crystal defects and the presence of grain boundaries and chemical impurities in the semiconductor material. The designer must always search for 'low noise' components for use in the low-signal parts of a system. While it is true that component manufacturers are continually offering devices with better noise performance, it is certainly the case that almost every instrument, and every part of it, generates noise in excess of the theoretical minimum.

Some of the sources of excess noise can be identified, while others are less clearly understood. However, many have in common a noise power spectrum that peaks at low frequencies, and that decreases approximately as the inverse of the frequency over quite a wide range. This noise is therefore not white, but 'pink', and is often referred to simply as *1/f noise* or as *flicker noise*.

A typical power density spectrum is illustrated in Fig. 5.2. We shall see shortly that this frequency dependence can be used to provide some discrimination against excess noise. For the time being, it is worth seeing how the 1/f noise affects experimental measurements. Very many measurements involve determining a current or voltage that is virtually at d.c. (zero frequency); of course, the signal has to change sometime to provide some information (see pp. 9–10), but the information is often conveyed by very low frequency components with periods of the order of minutes to hours. These are the very frequencies where the 1/f noise power density is maximum. The noise displays itself in the form of slow drifts of the measured quantity that may completely swamp the signal being sought. The solution must be to attempt to find a way of performing the experiment so that higher frequencies are used in the critical low-signal stages. We return to this topic later. The problems of noise and drift at very low frequencies is one reason why d.c. amplifiers are much more difficult to construct than a.c. ones. Small leakage currents, and slow fluctuations in them, that are present at the high sensitivity parts are amplified right through to the output. Although transistors do not show the same problems of ageing and heat generation as valves do, and although integrated circuits can employ many matched transistors in balanced configurations that offset the tendencies to drift, the fact remains that d.c. amplifiers are inherently more noisy (in terms of the noise power density near zero frequency) than are similar a.c. amplifiers.

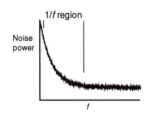

Fig. 5.2 Pink or 1/f noise

Interference

Noise arising from outside the experimental system is regarded as interference, and ought to be completely separable from the signal. However, a whole book would not suffice to describe the sources of interference and the cures, some rational and some verging on magic, that are employed. We must content ourselves here with a very brief survey. The electrical mains supply is one major contributor to interference. It is familiar enough in the form of 'hum' in audio equipment. Inadequately smoothed power supplies are one cause. Inductive or capacitative coupling to a.c. supplies internal to the equipment are another. Furthermore, the whole of the laboratory environment is subjected to alternating electric fields that can induce voltages virtually up to the mains supply voltage. Because motors, pumps, lights, heaters, and even scientific instruments, are all drawing current, an alternating magnetic field is present. These alternating fields can be prevented from entering the equipment only by adequate electrostatic and magnetic *shielding*. More insidiously, shielded cables that are grounded in more than one place may act as small transformers coupled into the oscillating magnetic field. The cables are then said to constitute a *ground loop* or *earth loop*. One characteristic of the mains *pick-up* in all these cases is that the frequency of the interference is that of the mains itself (50 Hz in the UK), and because the waveform is not very pure, harmonics of it. Other sources of a.c., such as local radio stations or r.f. power supplies in neighbouring rooms, can provide fixed frequency interference. Sometimes, inductive magnetic or capacitative electrical coupling is involved, but at higher frequencies a propagated electromagnetic wave may be detected by some part of the circuitry that is acting, possibly inadvertently, as a radio receiver. Similar remarks apply to pulse interference. Large amounts of power are dissipated suddenly by discharges in lasers and flash lamps. The outcome is a large pulse of energy that may be propagated directly as an electromagnetic wave or transmitted electrically down the mains supply wiring. The frequency spread associated with a short pulse is correspondingly large (see p. 7), so that this kind of noise can be nearly white. It occurs, however, at times associated with the pulses. Good detective work is often needed to track down what device in the laboratory is firing at particular intervals and at certain times of the day. Lasers are by no means the only culprits. Motors, thermostatically switched ovens, photocopiers, and almost all other electrical devices can be responsible. The counsel of perfection is to provide sufficient screening and shielding of the experiment and to filter the incoming mains supply adequately. It is often easier, scientifically if not socially, to eliminate the interference at source.

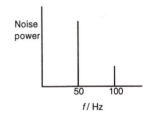

Fig. 5.3 Typical noise power spectrum of interference. The interference is often at the frequency of the mains and at harmonics of it

5.2 Signal-to-noise ratios and averages

Let us consider first the noise on a d.c. signal, which is the situation most often encountered at the output of an experiment (see p. 64). Figure 5.4 shows an average d.c. voltage level, and random noise superposed on it. A useful definition of *S/N* is then the ratio of the average signal to the root mean

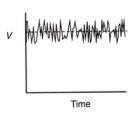

Fig. 5.4 Noise superimposed on a constant d.c. signal

Formula 5.3 Signal-to-noise ratio. For a hypothetical series of *n* individual signal voltages V_1, V_2.., the rms noise is given by the expression

$$(\overline{V_n^2})^{1/2} = \left(\frac{1}{n}\left[(V_s - V_1)^2 + (V_s - V_2)^2..\right]\right)^{1/2}$$

where V_s is the average signal:

$$V_s = \frac{1}{n}\left[V_1 + V_2...\right]$$

Thus

$$S/N = \frac{\text{average signal}}{\text{rms noise}}$$

$$= \frac{V_s}{(\overline{V_n^2})^{1/2}}$$

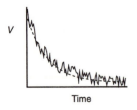

Time

Fig. 5.5 Noise superimposed on a decaying transient signal

square (rms) noise, as defined in formula 5.3. The rms is a convenient mean of the noise to use, because the squaring of the deviations means that both positive and negative excursions from the mean contribute to the total. For a varying or alternating signal, the definition of *S/N* is slightly more complicated, because the noise has to be referred to the expected true time-dependent signal, as illustrated in Fig. 5.5 for a decaying transient signal. Nevertheless, the same principles apply. It is worth noting that, if the noise stays at the same level while the signal decays, which would be the case for thermal and excess flicker noise, the *S/N* ratio degrades markedly with time, a factor that must be considered when weighting data points in subsequent analysis.

We must now examine what happens to the *S/N* ratio if two or more sets of measurements are averaged. The total signal, *S*, builds up in proportion to the number of readings taken and added together. The noise values do not add up in quite the same way. Sometimes the noise voltages add to reinforce each other, but sometimes they subtract from each other. The total noise thus increases less slowly than the signal, and the *S/N* ratio improves as more readings are taken to make up the average. It is easy to show, as is illustrated in Formula 5.4, that with *M* separate measurements averaged, the *S/N* ratio improves by a factor $M^{1/2}$. The central feature in demonstrating this result is that the cross terms in noise, such as $V_{n1}V_{n2}$, vanish *on average* if the noise is truly random, because there are as many random positive terms as negative ones. Only the squared terms remain, and the mean square of the total noise is equal to the sum of the individual mean square noise contributions.

The conclusion that the *S/N* ratio increases with the number of measurements taken is an extremely important one. In fact, the increase by the factor $M^{1/2}$ is exactly the same as ordinary simple statistics would predict for the increase in precision of the mean of any repeated determination, be it of a mass, a length, a temperature, or, as here, an electrical quantity. Many techniques used to enhance *S/N* ratios from a given noisy input signal depend, in one way or another, on the principle of making repeated measurements. Sometimes, the application of the principle is explicit as in the *boxcar integrator* (Section 5.5) or in a *multichannel averager* (Section 5.6). However, other methods such as those that rely directly on bandwidth reduction (Section 5.4) also make use of the cancellation of random noise. In every case, the *S/N* ratio improves in proportion to the square root of the time taken to make the set of repeated discrete measurements or taken to acquire the data. While it is fortunate that *S/N* ratios increase at all with extended observation times, the limitations imposed by the square root dependence should not be forgotten. Suppose that the *S/N* ratio is 1/10 for a one-second experiment; at this level, the signal is barely detectable above the noise. To obtain a precision in an experiment of ten per cent will require *S/N* to be roughly ten, which is 100 times larger than before. That means that 100^2 experiments must be done or an equivalent amount of information accumulated, taking (at least) 10^4 seconds, or rather under three hours. Three hours might be an acceptable time for a determination in a chemical experiment. However, to improve the precision to one per cent requires a further increase of a factor of ten in the *S/N* ratio, and thus of one hundred in the time taken to obtain the data. Now the experiment takes more than 11 days. It may prove impossible to operate an experiment satisfactorily for periods of this kind; reactants may run out, or factors such as temperatures, atmospheric pressures, and so on, may cause drifts or otherwise invalidate the experiment. There are thus practical limitations to the improvement of *S/N* ratios by any of these methods, and it remains important to design experiments to minimize noise at the outset, and not to waste signal (see Section 5.7).

5.3 Distinguishing characteristics of signal and noise

The improvement of *S/N* ratios depends on there being characteristics of the noise that differ from those of the signal. We have just seen that truly random noise tends to cancel over time. This randomness implies no preference for one frequency over another, and such random noise is white (see p. 62). Even pink noise (see p. 64) cancels over long enough averaging periods. The signal, on the other hand, occurs at or near d.c., or has well defined a.c. or time-dependent components. External interference (see p. 65) sometimes involves narrow-band a.c., but the frequencies are not related to those of the signal of interest: if they are, the experiment has been badly designed. One method of discriminating between signal and noise is thus based on the *frequency spectrum*, and Section 5.4 explores some of the possibilities.

Time may also be used to distinguish between signal and noise. Signals in some experiments may be anticipated only at certain times, whereas the random noise will be continuous in time. An example of such an experiment can be found the formation of an absorbing intermediate in a photochemical experiment initiated by a short laser pulse. It might be desired to obtain the spectrum of this intermediate. Figure 5.6 shows the time dependence of the concentration of the intermediate, called B here; B builds up to a maximum after the laser pulse has been fired and then decays. There is only any point in examining the absorption spectrum while substantial amounts of the intermediate remain, otherwise only noise, and no signal, will be present. The *S/N* ratio is enhanced by making use of the distinction in time between the signal and the noise. Section 5.5 considers how use can be made of this behaviour, particularly in cases where the experiment can be made repetitive.

Phase information is related to time for repetitive a.c. signals, and can be used as another method for distinguishing between signal and noise. Although not the most important benefit of the *phase-sensitive detector* (Section 5.4), the discrimination between signal and noise leads to an improvement in *S/N*.

5.4 Bandwidth reduction and integration

Low-pass filters and integrators

Starting, as before, with the idea of a signal that consists of a slowly varying d.c. level, one idea immediately presents itself for improving the *S/N* ratio.

Formula 5.4 Improvement of *S/N* ratio with increasing numbers of samples.

Consider first an average of two measurements in which the signal voltages are V_{s1} and V_{s2}, and the rms noise voltages are $(\overline{V_{n1}^2})^{1/2}$ and $(\overline{V_{n2}^2})^{1/2}$. The total signal is given by $V_s = V_{s1} + V_{s2} = 2V_{s1}$ since $V_{s1} = V_{s2}$. The mean square noise is

$$\overline{V_n^2} = (V_{n1} + V_{n2})^2$$
$$= \overline{V_{n1}^2} + \overline{V_{n2}^2} + 2\overline{V_{n1}V_{n2}}$$

But $V_{n1}V_{n2} = 0$ for random signals, so that

$$\overline{V_n^2} = \overline{V_{n1}^2} + \overline{V_{n2}^2} \approx 2\overline{V_{n1}^2}$$

for roughly equal noise on each signal. Thus,

$$S/N = \frac{2V_{s1}}{(2\overline{V_{n1}^2})^{1/2}} = 2^{1/2}\frac{V_{s1}}{(\overline{V_{n1}^2})^{1/2}}$$

In general, for *M* readings

$$S/N = \frac{MV_{s1}}{(M\overline{V_{n1}^2})^{1/2}} = M^{1/2}\frac{V_{s1}}{(\overline{V_{n1}^2})^{1/2}}$$

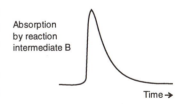

Laser pulse

Absorption by reaction intermediate B

Time →

Chemical reaction

$$A_2 + h\nu \rightarrow A + A$$

$$A \rightarrow B \rightarrow \text{products}$$

Fig. 5.6 Transient absorption in a chemical reaction initiated by a laser pulse

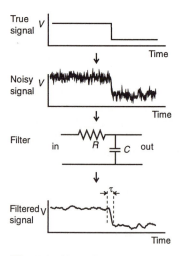

Fig. 5.7 Use of a low-pass filter to reduce high-frequency noise

Random noise has a power density spectrum that is independent of frequency, so that a large proportion of the noise is rejected by a filter that passes only d.c. and low frequencies. The *S/N* ratio has been enhanced because of the *bandwidth reduction* imposed by the filter, which passes only a more or less narrow band of frequencies, starting at zero frequency. A simple *low-pass filter* of the appropriate kind was introduced on p. 14, and the frequency response was shown in Fig. 1.20. Figure 5.7 illustrates how such a filter removes the high frequency components from a noisy signal. So far as the signal itself is concerned, the main consequence of adding the filter is that the abrupt change in d.c. level becomes distorted, the filtered output dropping to its new level with a time constant, τ, approximately equal to the product RC (see p. 14). The a.c. components are attenuated more and more as the cut-off frequency ($1/(2\pi RC) = 1/2\pi\tau$: see Fig. 1.20) is decreased, so that *S/N* increases with increasing RC. The rms noise voltages or currents for white noise are proportional to $(\Delta f)^{1/2}$ according to Formula 5.1 or 5.2, so that *S/N* ought to increase as $\tau^{1/2}$ in accord with the general principle established in Section 5.2. The maximum value of τ that is tolerable is determined by the demands of the experiment, including the length of time for which it will operate, as discussed on pp. 66–67, (and the patience of the experimenter). Filters can be designed (perhaps in the form of *active filters* that are based on amplifier-feedback circuits) that can have more satisfactory response characteristics compared with those shown in Fig. 1.20, such as a faster fall in response with increasing frequency beyond the 'knee'. Nevertheless, the trade-off between precision and response time remains.

In reality, another serious obstacle may exist to satisfactory improvement of the *S/N* ratio by filtering a d.c. signal. Examination of the noise imposed on the signal in Fig. 5.7 will show that the low-frequency excursions are relatively larger than the high-frequency ones; this excess low-frequency noise is exactly the situation represented by the pink noise of Fig. 5.2, and one that is very frequently encountered in practice. Most of the noise power is concentrated near zero frequency, so that attempts to remove the noise will also attenuate or distort unacceptably the signal changes that are to be measured. A solution to the problem will be examined in the next Section.

Integration is an alternative to the use of low-pass filters in conditioning noisy d.c. signals. Analogue circuits such as those of Fig. 4.15 (p. 58) may be used, or digital computer techniques (Chapter 6) may be harnessed. A linear integrator accumulates a running sum of the input voltages, and thus gives a constant weighting to all values as they contribute. Figure 5.8 illustrates the type of response expected for the noisy input of Fig. 5.7. The total integrated voltage increases with time, although the positive and negative noise excursions add and subtract from the running total. When the signal level changes, the underlying slope of the integrated output also changes. Dividing the voltage increase by constant time intervals gives the average signal over the integration period.

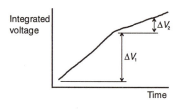

Fig. 5.8 Integrator output for the noisy input of Fig. 5.7

Modulation, chopping, and a.c. techniques

As emphasized in connection with the noisy signal of Fig. 5.7, the excess low-frequency components associated with many noise sources makes recovery of simple d.c. signals exceptionally difficult. One way of getting round this difficulty is to arrange for the signal itself *not* to be a slowly varying d.c. one, but rather an a.c. signal whose frequency is sufficiently high that it is clear of the obstructive $1/f$ region (see Fig. 5.2). Signals from many experiments are a.c. right from the outset, and so long as the frequencies are satisfactory, can be subjected to bandwidth reduction to obtain enhancements in *S/N* almost approaching the theoretical limit for white noise. For these cases, a *tuned band-pass filter* is used to limit the bandwidth. Filters based on combinations of high and low-pass units may be used, or resonant circuits may be employed (see p. 15 and Fig. 1.21). Although the applied a.c. signal level may change abruptly, the output of the filter takes some time to settle down. This behaviour of the band-pass filter is of fundamental origin. Any resonant system will exhibit damped oscillatory behaviour when subject to an impulse (including a change in sine wave amplitude). A wine glass or a tuning fork are familiar acoustic examples. The higher the Q of the system (corresponding to a narrow bandwidth: see p. 15), the longer this 'ringing' persists. There is thus a limitation in how rapidly a measurement can be taken, so that yet again the improved precision of the measurement given by the increased *S/N* forces an increase in the time needed to obtain the information, the time constant τ being $1/\pi\Delta f$ (this formula is similar to that for the low-pass filter for d.c., except that the latter has a factor of two in it because it is missing the contributions below the central frequency).

Improvement in the *S/N* ratio apply not only to the signal source itself, but also to the amplifiers used at small signal levels, because a.c. amplifiers are inherently less noisy and subject to drift than d.c. ones (see pp. 63–64 and also note comments on p. 55). To take advantage of the potential gains in *S/N* for signals that are slowly varying d.c., or at a.c. frequencies too low to clear the $1/f$ noise, it is necessary to convert the signal to a suitable a.c. frequency by a process of *modulation* or *chopping*. At the very simplest level, we might imagine switching a d.c. input signal at a constant frequency derived from an oscillator, as illustrated schematically in Fig. 5.9. With equal on and off periods, the output will be a square wave, the amplitude of which will follow the (slow) variations in the d.c. input level. The square wave can be amplified and processed as it is, or the fundamental frequency (see pp. 5–6) filtered out. Even this simple trick allows use of an a.c. amplifier, and forms the basis of the *chopper amplifier*. The switch itself is almost always an electronic one (although mechanical devices were used in the past). Probably, it will be desirable to convert the output signal after amplification back to d.c. for display or recording purposes. This *detection* or *demodulation* may be achieved with a straightforward rectifier circuit, such as the one shown in Fig. 1.25, accompanied by a suitable low-pass filter. However, it is virtually universal practice to employ, instead, a *phase-sensitive detector*, which will be described in more detail in the next Section.

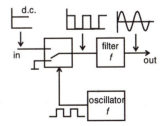

Fig. 5.9 Chopping of a d.c. input signal to generate a modulated a.c. signal

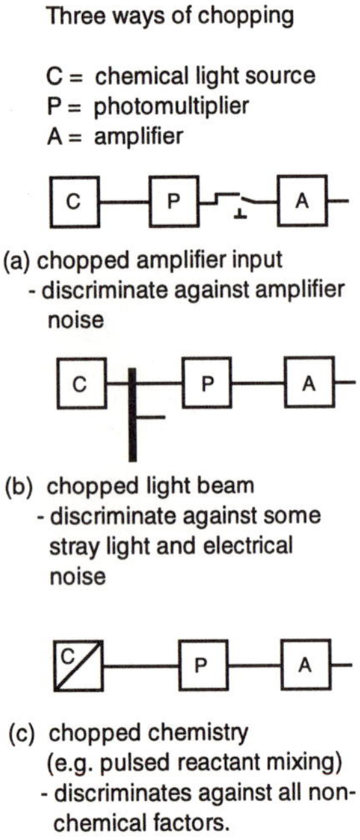

Three ways of chopping

C = chemical light source
P = photomultiplier
A = amplifier

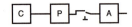

(a) chopped amplifier input
- discriminate against amplifier
noise

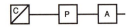

(b) chopped light beam
- discriminate against some
stray light and electrical
noise

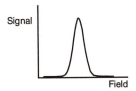

(c) chopped chemistry
(e.g. pulsed reactant mixing)
- discriminates against all non-
chemical factors.

Fig. 5.10 Three ways of chopping or modulating the electrical signal from an experiment

Signal | Field

Fig. 5.11 Absorption band in an NMR spectrum displayed as a quasi-d.c. signal as a function of field strength

Simple chopping of the d.c. input signal allows a.c. amplifiers to be used but any noise originally present on the signal itself will be amplified along with the signal. Chopper amplifiers are therefore only of great value when the signal is noise-free in comparison with the signal-processing electronics. In the more usual case, where the signal itself is noisy, some means has to be provided of modulating the phenomenon that produces the electrical signal in the first place. One common example of such modulation is the use of mechanical choppers in optical experiments. Figure 3.11 shows the principle with a sectored disc used as a rotating chopper wheel. Light from the source passes through the rotating blades, and the light falling on the photodetector is therefore 'switched' at a frequency determined by the rate of rotation and the number of blades. The chopped light will generate a (square-wave) alternating current in the photodetector, which can then be amplified using a.c. techniques. With a narrow-band a.c. filter, much low-frequency $1/f$ noise can be rejected. Furthermore, discrimination is provided against interference from stray light that falls directly on the detector, because it will not, in general, fluctuate at the same frequency as the chopper. Stray light that *does* pass through the blades is not rejected, so that an even better solution is to modulate the light source of interest itself. The feasibility of this suggestion depends on the nature of the experiment. Feeble light from a star can hardly be turned on and off in the source, but feeble light from a chemical reaction could be modulated by modulating the *chemistry*. The closer one gets to modulating the phenomenon of interest (the chemistry, here) rather than the effect it produces at later stages (first the light, then the electric current), the better chance there is of rejecting noise, as suggested by Fig. 5.10.

Other examples of modulation of the 'chemical' phenomena are found in many spectroscopic instruments, and especially in *magnetic resonance* spectrometers, such as an NMR machine. This is not the place to describe how such a spectrometer works in terms of the physical chemistry. Most readers will be familiar with the basic arrangement of a swept-field NMR spectrometer. The sample is held between the poles of a magnet (an electromagnet in our example), and radiofrequency (r.f.) energy is also supplied to the sample by means of a circuit that can also detect the absorption of energy. The principle is that the magnetic field is increased by changing the energizing current through the main magnet windings. When the nuclear spin resonance condition is met, absorption of the r.f. energy is detected. The display of absorption, in terms of voltage, against magnetic field (Fig. 5.11) is then the NMR spectrum. Except in very special cases, the signal would, in reality, be far too small to be detected satisfactorily for most applications in the way described. Instead, modulation is applied to the sample by using small auxiliary coils that are supplied with sine-wave a.c. at a frequency much higher than the sweep repetition rate (but much lower than the r.f. of the resonance signal). Figure 5.12 shows how the absorption signal responds. Off resonance there is no absorption, and therefore nothing to generate an output signal. On the approach to resonance (Fig. 5.12(a)), the absorption increases with increasing auxiliary field, and vice versa. The output signal is thus an a.c. voltage (superposed on a d.c. level) whose frequency is equal to the modulation frequency. At the peak of the absorption, the absorption does not change with field for small variations (Fig. 5.12(b)) and there is no output again. Figure 5.12(c) shows what happens on the other side of resonance. An a.c. signal is again obtained, but it now decreases as the field increases, so that the signal is *out of phase* with the modulation. In fact, the signal has an output that is proportional to the *slope* or *gradient* of the absorption spectrum of Fig. 5.11. If the a.c. signal is demodulated so as to preserve the phase information (see next section), an output such as that in Fig. 5.12(d) is obtained. The output is the *differential* or *derivative* of the true spectrum. The point provides an interesting illustration of the effects of two different kinds of modulation. In large amplitude modulation, such as with the light chopper, the entire 'effect' is turned on and off. One could envisage such a modulation for the NMR case if the magnets could be switched on or off rapidly, or subject to large amplitude modulation. The whole absorption would be in or out of resonance, and the modulated a.c. response voltage would be proportional to the absorption and not to its slope; that is, the integral spectrum would be obtained. Although it is not usual to switch magnetic fields in this way, the exact electrical analogue of this method is used in some *microwave absorption spectrometers*, in which a large square-wave electrical field modulates the absorption via the *Stark effect*.

Phase-sensitive detectors and lock-in amplifiers

The *phase-sensitive detector (psd)* plays a central part in instrumentation, because it provides a very satisfactory method of demodulating a.c. signals in an efficient manner. It is, indeed, phase sensitive, and can therefore be used to produce a demodulated output such as the one shown in Fig. 5.12(d). The term *lock-in amplifier* is usually applied to an electronic instrument or building block that contains both amplifier and phase-sensitive detector elements in it, and probably tuned filters and reference signal conditioning (see later) as well. Alternative terms that imply essentially the same as phase-sensitive detector are *synchronous*, *coherent*, or *homodyne* detector.

A simple diode detector (see p. 17 and Fig. 1.25) is a kind of switch, where the switching operation is initiated by the applied voltage itself. The diode and a imaginary switch equivalent are shown in Fig. 5.13. If the voltage goes positive, the switch is on; if it is negative, the switch is off. All voltages passed are positive, therefore, and the a.c. has been demodulated (half-wave rectified); after smoothing by a low-pass filter, a d.c. output is obtained (see Fig. 1.25 again). The output is always positive, and it is *not* sensitive to phase. A very simple modification to the switch circuit turns the detector into a phase-sensitive detector. All that is required is that the switching circuit is controlled *independently* of the signal by a *reference* waveform, as in Fig. 5.14. For our purposes, it is convenient to think of this waveform as a square wave, but any other regular waveform would do so long as it turns the switch on for positive half cycles and off for negative half cycles. In the examples of the chopped light beam that we have just

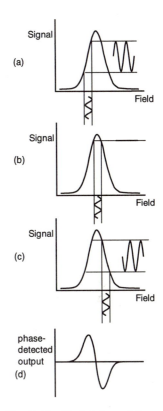

(a)

(b)

(c)

phase-detected output

(d)

Fig. 5.12 Modulation of the magnetic field in an NMR spectrometer

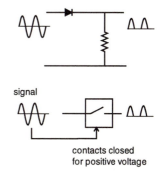

signal

contacts closed for positive voltage

Fig. 5.13 The diode as a switch

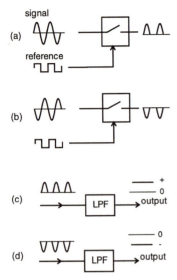

Fig. 5.14 A half-wave phase-sensitive detector. Output waveforms are given for signals (a) in-phase and (b) 180° out of phase with the reference. Addition of a low-pass filter (LPF) produces the smoothed outputs (c) from (a), and (d) from (b)

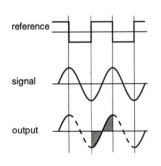

Fig. 5.15 Signal waveform that is 90° behind the reference

examined, this reference signal might be derived from a separate small lamp and photodiode combination placed on either side of the chopper disk. For the NMR spectrometer, the reference might be the oscillator itself that applies excitation to the modulation coils. Any signal (current produced in the main photodetector, NMR absorption voltage) will be locked in frequency to the reference waveform, but is not directly derived from it. Figure 5.14(a) shows what happens if a signal appears at the input to the switch if it is in phase with the switching reference waveform. Positive half cycles are passed. However, if the signal leads (or lags) the reference by half a cycle (180° out of phase), then the *negative* half cycles are passed, as illustrated in Fig. 5.14(b). After low-pass filtering, a positive or negative d.c. output is obtained (Fig. 5.14(c), (d)): the output of this detector is thus sensitive to phase. The demodulated output voltage is proportional to the rms a.c. input voltage. It should immediately be apparent how the use of a psd gives rise to the differential demodulated output shown in Fig. 5.12(d). For practical purposes, it may be more satisfactory to employ a *full-wave* psd, in which there are no half-cycle gaps between the switching periods, but we can illustrate the general behaviour of the psd in terms of the half-wave device.

We start by extending the previous discussion to the situation where there is a phase difference of 90° between signal and reference waveforms, as shown in Fig. 5.15 for a 90° lag. Assume that the switch is closed, and current flows, only when the reference signal is positive. The output consists of half cycles of the signal which have equal positive and negative contributions. After processing by a low-pass filter, the net output will be *zero*. From here, it is only a small step to imagine what happens as the phase difference between reference and signal waveforms is shifted by a full cycle. Starting with the two waveforms in phase, when the output takes its maximum positive value, the output drops with increasing phase difference to reach zero at a quarter of a cycle shift. The output becomes negative for larger phase differences, reaching its maximum negative value for one-half cycle shift. From that point on, the changes are reversed, the output becoming zero at three-quarters of a cycle, and fully positive at one full cycle.

In using a psd as a demodulator, it is normally required that the signal and reference waveforms be in phase to provide the largest possible positive output. Some electronic means of adjusting the relative phases is often provided by introducing an adjustable delay into the reference voltage.

When there is no fixed frequency or phase relationship between the sample and reference waveforms, the output always averages to zero if enough cycles are examined. For high-frequency signals, this property is obvious, as shown in Fig. 5.16(a) for a high-frequency noise waveform. Even within one half-cycle switching period, the negative and positive contributions to the net signal will come approximately to zero, and averaging with the signals in subsequent half cycles will bring the net output even closer to zero. With signal frequencies nearer to the reference frequency, the cancellation still occurs, but now over more cycles. Figures 5.16 (b) and (c) represent the situations for signal frequencies higher and lower than the reference frequency.

More than the sensitivity to phase, it is this ability of the psd to discriminate against all frequencies other than the reference frequency that constitutes the true power of the psd. The device acts, in fact, as a bandpass filter tuned to the reference frequency, and it can therefore reject noise by bandwidth limiting. The effective bandwidth is determined by the number of cycles over which the samples are averaged, and thus by the cut-off characteristics of the low-pass filter. The bandwidth at the a.c. frequency is closely related to the time constant or cut-off frequency of the band-pass filter used at d.c.. Since this time constant can be seconds or more, it is apparent that the bandwidth of the psd can be less than one hertz. Yet this bandwidth is imposed on the a.c. frequency, which might be hundreds to millions of hertz. The ratio of bandwidth to centre frequency is thus very large (corresponding to a very large effective Q factor), and the performance is difficult to obtain in other ways. Furthermore, the psd *tracks* the reference exactly, the centre of the pass band remaining locked to the reference frequency even if the latter wanders or is changed deliberately well outside the bandwidth of the filter itself. These qualities allow the psd to gain the maximum advantage from bandwidth narrowing of modulated signals, and permit the satisfactory recovery of weak signals in which the raw input starts off with a very unpromising *S/N* ratio. A further gain of a factor of two arises from the partial or complete rejection of signals of the correct frequency but wrong phase. It is no wonder that the psd finds such widespread application in so many kinds of instrumentation.

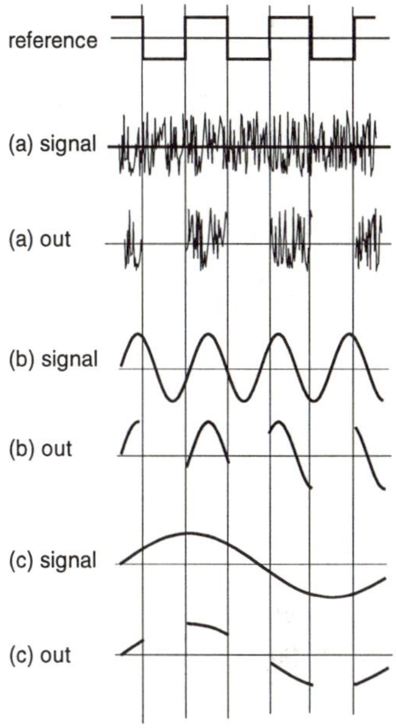

The psd shows its value most clearly when it is used to demodulate a repetitive alternating a.c. waveform. If the input signal is not a repetitive one, then some means of modulation must be provided as described. This theme of making a signal repetitive runs throughout the methods adopted to enhance *S/N* ratios. Another way of looking at the psd is to regard it as a *signal averager*, the average being taken of the individual half-cycle waveforms over the time constant of the low-pass filter. As emphasized on pp. 65–67, signal averaging and bandwidth reduction are different facets of the same process.

Fig. 5.16 Demodulation of waveforms not related to the frequency of the reference

5.5 Time discrimination and averaging

The output from the psd after filtering gives information about the average (rms) voltage of the a.c. and about its phase, but not about the shape of the waveform. In this section, we discuss the *boxcar detector* or *integrator*, which is able to extract the *wave shape* by a process of analogue signal averaging of a repetitive signal. The device is also able to provide discrimination between signal and noise on the basis of the time of occurrence, as described in Section 5.4, and it is this aspect that we shall examine first.

Refer again to Fig. 5.6, which shows a signal pulse (optical absorption of a reaction intermediate is the example used) that follows a triggering signal (a laser flash in the example), and grows and decays (with characteristics determined by the chemistry). This experiment can be made repetitive just by retriggering the laser at regular intervals. The repetitive waveform is

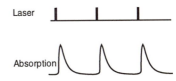

Fig. 5.17 Repetitive pulsed experiment (Fig. 5.6 redrawn for repeated laser shots)

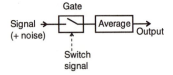

Fig. 5.18. A switch and averager ('sample-and-hold' circuit). Note that a gate is said to be open (allows current to flow) when the switch is closed

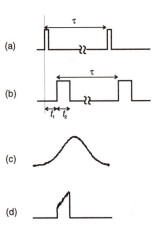

Fig. 5.19 The boxcar integrator: (a) trigger, (b) sample gating signals; and (c) input, (d) sampled waveforms

represented in Fig. 5.17. Only noise, and not signal, is present during most of the period between laser pulses, a situation in which the signal is said to possess a small *mark-to-space* ratio (in contrast to the modulated signals of the last section, which were presumed to have unity mark-to-space ratios). As suggested in Section 5.4, it is wasteful of the *S/N* ratio to accumulate data throughout the period between the laser pulses, because only noise is present during most of this period (the 'space'). A method of measuring only when the signal is present (the 'mark') will obviously be advantageous. The boxcar detector provides one method of making measurements in this way.

In essence, the boxcar detector is very simple, and consists of a switch or *gate* connected between the source of signal (and noise) and an averaging device, as indicated in Fig. 5.18 (see the note about open and closed gates). In its simplest form, the averager might just be a low-pass filter with a time constant much longer than the period τ between repetitions. The generation of the signal that activates the switch is of central importance. What is required is a signal that will turn the switch on at a fixed period, t_1, after a trigger pulse from the laser, and hold it open for another fixed period, t_2. The waveform needed (b), and its relation to the laser trigger (a), is shown in Fig. 5.19. A possible way of achieving the result is shown schematically in Fig. 5.20. It is the shape of the gating signal of Fig. 5.19(b) that gives rise to the boxcar's name, because the shape is said to resemble that of a railway wagon (which, in America, is called a boxcar). Now we examine how the averager responds to the noisy absorption signal. Figure 5.19(c) shows the absorption in a single shot on a time axis expanded from that of Fig. 5.17. This signal reaches the averager only during the period t_2, and after the delay t_1. After a time τ (see Fig. 5.19), the entire sequence repeats itself as the laser fires another shot. Typically, the period τ is much longer than either t_1 or t_2. At each shot, the gate is opened, and because of the long time constant of the low-pass filter, the final output is the average of the signals that are present during many repetitions. These signals are, in the system we have described, made up of a noise-free zero voltage when the gate is closed, and the signal + noise when the gate is open, equivalent to a simple scaling factor (t_2/τ) multiplied by the average of signal + noise during the period t_2. The final output is thus effectively the result of averaging many gate-open samples, and the *S/N* ratio will be improved over the single-shot value by the square root of the number of cycles averaged (or by the equivalent bandwidth reduction). This is the usual gain expected from frequency discrimination. In addition, since noise is present only during the period t_2, the *S/N* ratio is also improved over that for an average obtained without time discrimination (that is, without gating) by the factor τ/t_2.

One disadvantage of the device as it has been described is that for very short gate periods and long repetition periods, the ratio t_2/τ may make the output voltage very small. Modern implementations of the boxcar integrator often employ *sample-and-hold* circuits ahead of the final low-pass filter that *sample* the signal while the switch is closed, and *hold* the information when the switch is open. In this way, the averages of each individual gate-open voltage are themselves averaged, but the long zero periods between repetitions are not included. Repetitive waveforms lasting for a few picoseconds, but repeated at intervals from microseconds to much longer, can be studied successfully with this kind of instrumentation. One of the necessary skills in using the boxcar is in choosing the gate period and delay to exclude the unwanted noise, yet accept almost all of the wanted signal.

A comparison of the descriptions of the phase-sensitive detector and the boxcar detector as given here should indicate the close similarities of the two types of device. Both possess switches or gates driven by a reference waveform, and the gated signal (and noise) is averaged in some way with a low-pass filter. The reference waveform in the boxcar can be shifted to give an adjustable time delay from a trigger signal, and a similar delay *may* be introduced in the psd to provide electronic phase shifts. The difference is essentially in the shape of the gating waveform. In the psd, the mark-to-space ratio is unity, while in the boxcar it is (much) less than unity.

So far, we have considered the use of the boxcar integrator in recovering the amplitude of the signal present at a fixed delay t_1 after the trigger, with the objective of rejecting noise present at other times. The other use of the boxcar is in recovering the waveform itself when the signal is obscured by noise. Once again, it is the repetitive nature of the experiment that makes possible the improvement in the *S/N* ratio. The only difference in the experiment is that instead of using a fixed delay t_1, the delay is varied to encompass the entire waveform. The experiment could be done in a step-by-step manner, with the investigator adjusting the delay, and taking the reading of signal amplitude with an output time constant that is long enough to give the required S/N ratio. The delay is then set to a new value, and the signal voltage measured again. Ultimately, the complete waveform will have been explored, and the results can be plotted out by hand or otherwise. Note that improved precision on the amplitudes requires longer to be spent making each individual reading. Higher time resolution requires the gate-open period, t_2, to be reduced. More readings are therefore needed to cover the complete waveform; the experiment takes longer for a fixed precision in amplitude.

In reality, the manual point-by-point determination of a waveform is replaced by electronic scanning . Figure 5.21 shows how this scanning might be achieved, with the final result being displayed on a chart recorder. A slowly increasing sweep voltage is used both to vary the delay time t_1 and to drive the *X* axis of an *X–Y* chart recorder (see p. 52). The output of the boxcar averager itself drives the *Y* axis. As the delay is advanced, the pen moves along the horizontal axis, while the boxcar determines the averaged (and therefore *S/N*-enhanced) amplitude of the waveform at each delay. One aspect of the boxcar is thus that it apparently transforms the timescale over which the waveform is available for examination. The boxcar arrangement of Fig. 5.21 is a form of *transient recorder*. It may take many minutes for the sweep to be completed, during which time the waveform can have repeated itself many hundreds of millions of times.

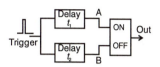

Fig. 5.20 A method for obtaining the timed gating signals for the boxcar integrator. A trigger pulse sets off two timed delays. The output pulse A from the first delay circuit opens the gate, while the pulse from delay B closes it again

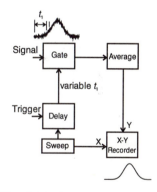

Fig. 5.21 A boxcar integrator with electronically swept delay for waveform examination

5.6 Multichannel methods

Multichannel transient recorders

The explanation of the mode of action of the boxcar integrator illustrates one of the problems associated with instrumentation of this type. Even if the waveform is investigated with a scanning electronic sweep, the process of recording it can be regarded as a step-by-step examination of successive portions of the waveform, each one separated by the gate width, t_2. Although the entire waveform could occupy a maximum time τ, the boxcar only collects information within the much smaller gate-width period, and all other information during the much longer period $(\tau - t_2)$ is wasted. If the initial *S/N* ratio is sufficiently large that the entire waveform can be built up with adequate freedom from noise within a total accumulation period of seconds to minutes, this waste may not matter much. However, if the *S/N* enhancement requires many repeated shots at each point, then the experiment may become impossible because it takes too long (either for the operator's convenience or, as we have emphasized earlier, for the reason that the experiment may not be capable of yielding unchanging data over the entire period). Consider a specific example, in which the waveform of interest spans a period of 1 ms, and the gate width (corresponding to the resolution on the time axis) is 1 μs. There are thus 1000 consecutive portions or 'channels' of waveform to be examined. Suppose further that the *S/N* ratio on the input waveform from a single shot is 1/10 and that it is required that this ratio be improved in the output to 10/1. The required enhancement is thus a factor of 100, so that, in theory, a total of $100^2 = 10^4$ individual waveforms must be averaged. The minimum time for the accumulation is thus this number multiplied by the repetition period of the experiment, which is at least 1 ms (but might be much more if the experiment cannot be repeated immediately after the output of interest has died away). Ten seconds, at least, must therefore be spent in obtaining the data. This length of time might be acceptable if the amplitude at just one time delay were required. However, to accumulate the amplitude data for all 1000 channels will require 10×1000 seconds, or nearly three hours, which may prove impossibly long.

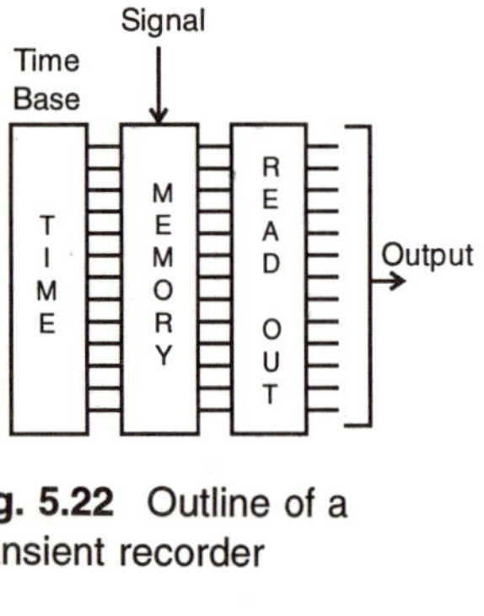

An enormous saving in time could obviously be obtained if *all* the channels could be recorded on *every* shot. In the example presented here, the gain would be the factor of 1000 that results from the number of hypothetical channels. It is exactly this gain that is provided by a series of *multichannel* devices. A *transient recorder* is an instrument capable of memorizing the amplitude–time profile of a single shot of a transient waveform. Transient recorders are manufactured with typically anything from a few hundred to a few thousand discrete channels in which the data are stored. The basic operation of the instrument is illustrated in Fig. 5.22. An internal *time base* or sweep generator (see pp. 26–27) is triggered by some event linked to the start of the transient waveform to be examined. The voltage ramp of the time base is translated to provide the sequence number of a memory channel. Into this channel is stored the amplitude information appropriate to the particular

Fig. 5.22 Outline of a transient recorder

time delay since the sweep started. The time per channel can be less than one nanosecond in the more sophisticated of the transient recorders available; there is no theoretical limitation on the maximum period. The data stored in the memories can be read out at any later stage by addressing each of the channels at a rate convenient for the application in question. The data might be used repeatedly to produce a continuous display on an oscilloscope, in which case the transient recorder has been used as an essential component of a storage oscilloscope (see pp. 27–28), or the data might be read out for plotting or subsequent processing in some form of computer. Indeed, the next stage in multichannel signal processing involves just such an addition of suitable numerical data handling that allows the averages of the waveforms of multiple shots to be calculated, more or less in real time. The result is sometimes called a *computer of average transients* or *CAT*, and is the multichannel equivalent of the boxcar integrator. The provision of many channels brings with it the so-called *multiplex advantage*, which means that for M channels, the same information can be obtained in $1/M$ of the time as in a single channel. For the *same* data acquisition time as the single-channel device, M times more information is obtained, which implies that the *S/N* ratio is improved by a further factor of $M^{1/2}$.

One could envisage one form of CAT as consisting of a whole array of hundreds or even thousands of boxcar integrators. Some of the very fastest transient recorders do, in fact, utilize something along these lines at the first stages of data capture. However, the last paragraph will suggest that digital techniques and memories might be particularly suited to the task of multichannel recording. The basic instrument is the *multichannel scaler*, which in its simplest form allows counts to be accumulated in memory channels. For the purposes of recording transient phenomena, these channels are addressed sequentially by the triggered time base. In this form, the multichannel scaler is, as its name implies, a *counting* device. For some sorts of instrumentation, where the signal appears as digital pulses (for example, from electron multipliers in mass spectrometers or from photomultipliers in low light-level experiments), this counting capability is exactly what is needed. When an analogue signal is to be averaged, which is the situation that we have envisaged hitherto in this section, it must first be digitized by use of an analogue-to-digital converter (ADC: see p. 22).

Multiplex spectrometers

Many scanning instruments, such as spectrometers, are essentially sequential in the manner in which they record information, and thus suffer from wastage in the same way as does the boxcar. Figure 5.23 shows the essential elements of a simple spectrometer for the visible or ultraviolet regions of the spectrum. The feature that should be emphasized here is the single exit slit that allows one band of wavelengths to pass through to the detector. To scan the spectrum, the grating angle is changed. Intensities are recorded at each wavelength (properly speaking, band of wavelengths) in turn. For

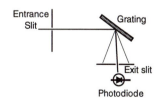

Fig. 5.23 Principle of a scanning optical spectrometer

low-intensity sources of radiation, noise is often a problem, and one of the methods described earlier in this chapter must be used to provide *S/N* enhancement at the expense of increased data acquisition time. If just one or a few wavelengths need to be examined, there may be no problem (compare this situation with the boxcar used to probe amplitudes at one delay time). However, if the entire spectrum must be examined, it is evident that there would be a considerable advantage in finding a multiplex alternative to scanning that would be analogous to the CAT or multichannel scaler. In the past, photographic plates in the focal plane of the spectrometer were used instead of the slit and photoelectric detector of Fig. 5.23. Although there are problems about convenience, sensitivity, and most particularly quantitative measurement using the photographic method, at least it does have the advantage of probing the whole spectrum simultaneously: that is, it is a multiplex method. Modern technology has brought a comparable photoelectric technique in which a *diode array* (see pp. 45–46) in the focal plane (Fig. 5.24) replaces the single photodiode of Fig. 5.23. The array is a specialized integrated circuit in which a large number of photosensitive elements are fabricated side-by-side on a single semiconductor wafer (usually silicon). Typically, there might be 512 or 1024 elements, each of which defines a small portion of the dispersed spectral image and thus corresponds to a small wavelength band. The outputs from the individual photodiodes are fed, often via sample-and-hold circuitry (see pp. 74–75), to its own memory that accumulates or averages the total photocurrent generated by that diode. Small 'personal' computers have proved ideal for the purposes of acquisition and subsequent numerical manipulation of the data (see also Chapter 6). Just as with the CAT or multichannel scaler (see pp. 76–77), the multiplex advantage for an *M*-element array brings a gain in speed of a factor *M* for the same *S/N*, or an improvement in *S/N* of $M^{1/2}$ for the same acquisition time, over the single-channel spectrometer.

Other examples of multichannel multiplex instrumentation in chemistry include devices that measure ejected electron energies in *photoelectron spectroscopy* and in *ESCA* (*electron spectroscopy for chemical analysis*). The information of interest is contained in the energy or velocity of the electrons that are removed from the parent species by UV or X-ray ionization. Normal scanning instruments probe the electron energy by use of an energy filter that transmits only electrons of the selected energy. There is thus the usual wastage of information that is possessed by the electrons that are rejected. One multiplex method allows the ejected electrons to pass through a perpendicular electric field, which bends them to an extent dependent on the electron velocity. The electrons are dispersed according to energy rather as the photons in the spectrometers of Figs 5.23 and 5.24 are dispersed by the grating according to wavelength. The electrons are allowed to fall on a fluorescent screen whose light is coupled into a diode array photodetector. Position along this detector therefore corresponds to electron energy, which is recorded for *all* electron energies simultaneously.

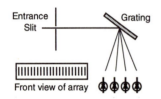

Fig. 5.24 Replacement of the slit and single photodiode of Fig. 5.23 by a diode array

Pulse height analysis

The *pulse height analyser (PHA)* is another multichannel instrument. One important application of pulse height analysis is in measuring the energy spectrum of ionizing radiation (e.g., of X-rays or gamma rays), and in nuclear instrumentation generally. Two examples of the use of this instrumentation in chemical analysis can be used to illustrate the use of the PHA. The first is *neutron activation analysis*, in which a sample is bombarded with neutrons so that some radioactive isotopes of the elements present are generated. If these isotopes produce gamma rays in their decay, the characteristic energy spectrum of the gamma rays can be used to identify the elements present with extremely high sensitivity. *Proton-induced X-ray emission (PIXE) spectroscopy* is similar in concept. Here, high energy protons are used to bombard the sample, and the energy of the X-rays produced is analysed, once again producing chemical identification of trace elements, but this time with the possibility of providing spatial resolution.

The determination of the energies of the radiation produced starts with the detectors employed. These detectors include gas-filled *proportional detectors*, *scintillators* in which the incoming radiation is converted by a suitable crystal or liquid to a flash of light that is measured with a photomultiplier, and a variety of solid-state detectors. All have the property of producing an output pulse for each incoming X-ray or gamma-ray photon that is proportional in amplitude to the energy of that photon.

The next step is to build up an energy spectrum of the ionizing radiation, and it is for this purpose that the PHA is used. Instead of directing information to a specific channel of a multichannel recorder according to the time at which the signal arises, as in the multichannel scaler, the choice of channel is made according to the amplitude of a pulse (see Fig. 5.25). The idea is to accumulate in a specific channel all input pulses whose amplitudes lie between certain chosen limits. The readout from such a device is then a histogram of the total number of pulses as a function of their amplitudes. Several straightforward electronic techniques are available to act as *lower* and *upper threshold discriminators* and the combination of a pair of such devices will constitute a *window discriminator*. Each incoming pulse is analysed for amplitude with a window discriminator, and the memory channel appropriate to that pulse height is incremented by one unit. The technique possesses the usual advantages of multichannel operation, and the signal-to-noise ratio increases as before in proportion to the square root of the number of counts accumulated in each channel. One constraint is that individual pulses must not arrive more rapidly than the rate at which their heights can be analysed and the appropriate memory addressed and incremented.

Time to amplitude conversion (TAC)

One further application of the PHA merits brief discussion here. Some measurements in chemistry involve very short time periods. For example, the decay of fluorescence may occur over a timescale of nanoseconds or less.

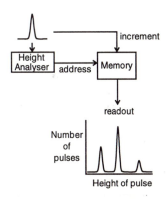

Fig. 5.25 The pulse-height analyser (PHA). The analyser (a specialized ADC) measures the pulse height and selects the appropriate channel of the memory. The contents of the memory is incremented by one count

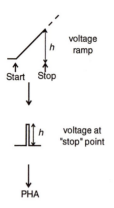

Fig. 5.26 Time-to-amplitude converter. A voltage ramp is started and stopped by the timing signals; the amplitude of the voltage at the end gives a measure of the elapsed interval, and can be stored in a pulse-height analyser.

The reader is advised to review the material on Fourier transforms presented on p. 9

The addressing in conventional multiscalers, in which the channels are selected sequentially as a function of time, is not fast enough to deal with these short periods. One alternative is to use a *time-to-amplitude (TAC)* converter in conjunction with a pulse height analyser. In the TAC, a voltage ramp is generated from the instant of arrival of a 'start' pulse until the arrival of a 'stop' pulse, as indicated in Fig. 5.26. The voltage that the ramp has reached is thus proportional to the time between the two pulses. This voltage is next converted to a single pulse of the same amplitude, and stored in a PHA. The rate of increase of voltage can be sufficiently large that time resolutions of the order of picoseconds are attainable. In the fluorescence experiment, the 'start' pulse might be derived from a flash of light used to excite the fluorescence and the 'stop' pulse from the first photon of fluorescent radiation detected by a photomultiplier. Each time the flash is triggered, the sequence repeats, and a new pulse height is stored in the PHA. Ultimately, the pulse-height 'spectrum' will correspond to the averaged distribution in time of the photons. While very fast, the TAC is only suitable for experiments that will provide just one 'stop' pulse for each ramp started. For the example of fluorescence lifetimes used here, this requirement translates to one fluorescent photon *or less* being detected for each flash of the exciting radiation (the TAC voltage ramp can be reset to zero if no photon is detected). A low-intensity light source is therefore needed, but the flash can be repeated as fast as the PHA can cope (10 000 times a second is typical in this kind of experiment) in order to build up a statistically significant picture of the time distribution of fluorescent photons with an adequate signal-to-noise ratio.

Fourier transform methods: a second look

The use in a multiplex instrument of a suitable number of channels, each of which replicates the basic measuring device, is conceptually straightforward. Arrays of boxcar integrators, counters, or photodiodes are all of this multiple detector kind. This solution to the improvement of *S/N* figures cannot be adopted universally, because of geometrical or cost problems, or just because of the sheer number of individual detectors that would be needed to provide the required range and resolution. For example, in NMR spectroscopy upwards of 50 000 individual narrow-band r.f. detectors would be needed. There is sometimes another way of getting at the multiplex advantage that relies on the use of Fourier transforms (see p. 9). The availability of computers allows mathematical computation of the transforms so rapidly that there is no significant obstacle to working in one Fourier domain, such as time, when what is required is information in the partner domain, such as frequency. Data may be collected in whichever domain is more accessible and convenient, and sometimes this procedure also brings with it the multiplex advantage. In chemical instrumentation, the use of Fourier transform techniques has become established for a variety of spectroscopies, including NMR, ion cyclotron resonance, and infrared, as well as for the study of other electrochemical, dielectric, and optical phenomena. In the present section, we review the principles by considering the analysis of the frequency spectrum of a complex alternating current waveform. Although the a.c. *spectrum analyser* is an instrument of particular value in investigating the behaviour of electronic devices, it is also an important building block or diagnostic device in several pieces of chemical instrumentation. The objective is to generate spectra, such as those in Figs 1.6(c) or 1.7(b) from the corresponding waveforms.

A simplified outline of a scanning spectrum analyser is shown in Fig. 5.27. Some frequency-selective device, which we have called a variable tuned filter in the figure, selects a narrow band of the a.c. and the output is converted to d.c. in the detector so that the amplitude can be measured. The filter is now tuned to a new frequency, and the amplitude measurement repeated. The process is continued until the whole spectrum has been examined. The production of a spectrum is entirely analogous to that in the optical case (compare with Fig. 5.23). To make the analyser a little more sophisticated, we have allowed the filter to be voltage controlled, with a sweep generator altering the bandpass frequency as well as driving the *X*-plates of an oscilloscope used in *X*–*Y* mode. The spectrum is thus displayed directly on the oscilloscope. For very high frequencies, this kind of arrangement continues to be used, but it suffers from the drawback, especially for noisy signals, that only one band is examined at a time, so that it may take excessively long to build up a sufficiently noise-free spectrum. One alternative would be to construct a multiple detector analyser, according to the outline in Fig. 5.28. This solution might be suitable for analysing a few rather wide bands in order to get an idea of the distribution of frequencies present, or for looking at a few selected spot frequencies. But for many channels, it becomes impossibly unwieldy, as well as prohibitively expensive.

Fourier transform techniques provide the key to multiplex spectrum analysis of a.c. at frequencies up to about 1 MHz (the limit is set by the speed of the digital devices that are used). The point to be remembered is that the waveform is the amplitude of the a.c. signal as a function of *time*, while what is required is the amplitude of the different components as a function of *frequency*. Since time and frequency are two domains related by the Fourier transform, it follows that the waveform can be recorded and the Fourier transform operation performed on it mathematically in order to yield the required frequency spectrum. Virtually all Fourier transform (FT) analysers and spectrometers now use digital techniques, in order that the subsequent processing can be performed directly by computer, which in turn almost always uses a discrete version of the (continuous) FT embodied in the *Fast Fourier Transform* (*FFT*). Figure 5.29 shows a block outline of an FT spectrum analyser. It should be compared with the transient recorder of Fig. 5.22. The essential difference is the addition of the FT computer after the memory. The description of digitization, timing, and storage of the signal given in connection with Fig. 5.22 applies equally to the FT spectrum analyser. The output of the FT computer is the frequency spectrum required. This output is, of course, itself digital and can be stored or manipulated accordingly, or it can be converted back to an analogue signal for display on

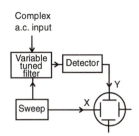

Fig. 5.27 A sequential (swept) frequency analyser

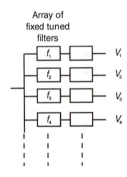

Fig. 5.28 Conceptual outline of a multiple-filter frequency analyser

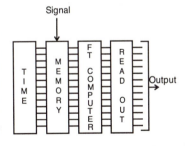

Fig. 5.29 Block diagram of a Fourier transform frequency analyser. Compare this diagram with Fig. 5.22. The essential difference is the addition of a computer

an oscilloscope. In a single shot of the waveform, the entire spectrum is built up, so that the potentiality of gaining the multiplex enhancement of the *S/N* ratio is provided. The multiplex information was obtained in time (by not allowing any part of the waveform to pass unrecorded) but it appears in the FT partner domain, the frequency. Of course, the possibility also exists of co-adding or averaging many waveforms if the signal is repetitive (as we have been assuming here). The expected further improvement in *S/N* may then be obtained. In fact, it is unimportant whether the averaging is performed on the waveform (time domain) or spectrum (frequency domain).

The number of independent spectral (frequency) data points or equivalent channels, *M*, that can be computed has a direct relationship to the duration over which waveform samples taken, so that the spectral resolution depends on the length of the sample. There are other sampling conditions that must be fulfilled as well, but these questions are ones of detail.

5.7 Limitations on the improvement of *S/N* ratios

This chapter has been concerned with the methods available for the recovery of signals in the presence of noise. Success in scientific instrumentation often turns around this problem, and the length of the chapter reflects the importance of the topic. A recurring theme has been that repetition of the measurements may lead to an enhancement of the *S/N*, corresponding to an increase in precision of the overall measurement, even though that improvement is at the expense of a longer time needed to obtain the result. Multiplex methods, including FT techniques, prevent the unnecessary waste of signal or time. We have sought to show that multiplex methods possessing *M* actual or equivalent channels should provide a gain in *S/N* of $M^{1/2}$ over the corresponding single-channel device. However, this maximum gain in *S/N* applies only to certain types of noise. In the first place, it has been assumed that the noise is truly white or random. If that is not the case, then the simple statistical arguments will not hold. With some kinds of $1/f$ noise, for example, the noise builds up in proportion to the signal, so that signal averaging offers no advantage, and FT methods can even degrade the *S/N* ratio, rather than improve it. In many cases, however, the noise is essentially white, and independent of the signal level. Thermal noise generated in detectors behaves in this way, and the noise is said to be *detector limited*. Both multiple detector and FT multiplex techniques provide the gain in *S/N* of $M^{1/2}$ over the single-channel detector. If the dominant form of noise is shot noise-on-signal (*source limited noise*: see pp. 62–63), then the advantages to be gained from multiplex experiments are different again. For the multiple detector experiment, the gain is $M^{1/2}$ as anticipated. For the FT experiment, however, there is no gain at all over the single channel method. It should be apparent, therefore, that a sensible approach to successful noise reduction really does require an understanding of the origins and characteristics of the noise, and that the brute-force application of the most sophisticated techniques may be at best a waste of time and at worst an unjustifiable squandering of resources.

6 Computers in instrumentation

Computer techniques are now used widely in instrumentation and we present here a brief survey in order to convey the just the basic concepts. Many of the instruments described in previous sections make extensive use of digital electronic techniques, with integrated circuits often employed to provide switching and logic functions as described at the end of Chapter 1 and the beginning of Chapter 2. The gating signals for the phase-sensitive detector (pp. 71–73) or the boxcar integrator (pp. 73–75) are, for example, very conveniently generated by suitable combinations of logic gates in which the timing is controlled by the external reference signal and, in the case of the boxcar integrator, by an internal clock. Multichannel recorders and scalers (pp. 76–77) require (digital) memory as well as the timing and addressing circuitry, and may also need ADC and DAC devices to interconvert analogue and digital signals at the input and output (p. 22).

Digital integrated circuits wired together in an appropriate way can provide sequences of logical operations of great complexity. However, except for alterations in function that can be provided by switches on the instrument, the operations themselves are predetermined by the circuits and the wiring. It is in this respect that the *computer* differs most significantly from the *hard-wired* circuits envisaged so far. Although the computer consists of logic circuits wired together (the *hardware*) to provide a relatively small number of more sophisticated operations, the order and repetition of the operations are under the control of a *program* (the *software*). The program may be modified by the designer or user of the instrument to provide different results without any change to the initial wiring of the instrument. Great versatility can therefore be achieved, and several tasks can be run in sequence so rapidly that they seem to be simultaneous. The breakthrough in the use of computing techniques has come with the development of *microprocessors* (see pp. 18–19). Only the largest experiments justified the use of a dedicated main-frame computer, which might occupy a large room and cost hundreds of thousands of pounds. A microprocessor can be contained in a package of area of a few square centimetres containing an even smaller silicon chip and cost as little as a few pounds. Both the size and cost of the integrated circuit microprocessors makes their incorporation into relatively simple instrumentation a feasible proposition. In general, such instrumentation uses not only the central processing unit (CPU) of the microprocessor, but also associated memories for both program and data, input and output devices, and possibly external storage media, so that the combination approaches the configuration of a *microcomputer* such as a 'personal computer' (PC). Indeed, commercial microcomputers are often used as packages instead of the individual component parts put together by the instrument manufacturer.

6.1 Applications of microprocessors in instrumentation

There are at least four ways in which microcomputers can be used in instrumentation. They are in

(i)   data processing;
(ii) data acquisition;
(iii) experiment management; and
(iv) feedback control.

We now consider these four areas in turn.

Main-frame computers were seen mainly as an aid to data processing. The underlying operations in logic of a programmable computer are easily translated to very rapid arithmetic and, by the power of repetition and speed, to complicated numerical mathematics. Fourier transform methods, for example, as outlined on p. 9 and pp. 80–82, were possible using main-frame computers in this way, but with the data usually recorded on the instrument and the computation being performed remotely and 'off line'. The dedicated microprocessor can, in contrast, carry out such number-crunching activities 'on line'. Although this aspect is one of great convenience, it is possibly one of the less important of the advantages that a dedicated microprocessor can bring to instrumentation.

Acquisition of data under computer control provides a general and versatile method for recording data, either directly as digital information, as in a computer-based multiscaler (p. 77), or after conversion with an ADC from an analogue input, as in the transient recorders (pp. 77–78) or the digital storage oscilloscope (pp. 27–28). Information about the timing of events may also be recorded. The data so obtained may be further processed numerically straight away, and it may also be stored (often on magnetic disk or tape) for subsequent recall and manipulation.

The use of computers to run instrumentation and experiments relies on the speed with which tasks can be performed in an error-free manner. Sequences of operations at predetermined times, such as opening valves, turning on lamps and power supplies, changing spectrometer slit widths, and so on, can be carried out in the right order and virtually as fast as desired. The outcome is that entire new classes of operations can be undertaken once the limitations of human accuracy, speed, and attention span have been removed.

Of the four uses of computers in instrumentation, the incorporation of microprocessors in control loops (see Chapter 4) is one of the most fascinating and advantageous. The essence of this application is that the control law (p. 60) can be adjusted intelligently to suit different circumstances, and the possibility arises of adaptive control systems that can optimize themselves. This application of microprocessors is, perhaps, less obvious than the other three, and is singled out for further exploration in Section 6.3.

6.2 Integrated circuit microprocessors

Some insight into the applications of microprocessors in instrumentation can be gained from a modest understanding of their internal structure. Figure 6.1 shows in schematic form the essential sub-components of a typical microprocessor. An internal *data bus* links the various parts, and also conveys data to the external environment. Different operations within the

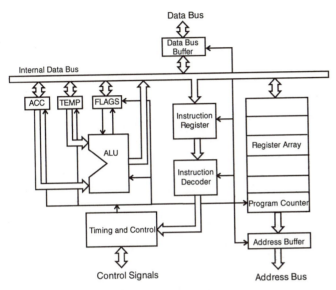

Fig. 6.1   Essential elements of a microcomputer

microprocessor locate a specific element of external memory, called an *address*, which is linked to the microprocessor by an *address bus*. The operations include the step reached within the external program as determined by the *program counter*, and addresses where the results of operations are to be stored or where data required may be found. The data bus in many modern microprocessors consists of sixteen parallel lines, so that the data may be transferred to and from memory sixteen bits (two bytes — see p. 21) at a time. Older, or simpler, microprocessors relied on data busses eight, or even just four, bits 'wide'. The main difference is that a number consisting of many digits (such as a large integer or a high-precision floating-point number) requires more individual transfer operations, and more complex mathematical operations, for the narrower busses. The sixteen-bit bus device, for example, can calculate or transfer 2^{16} positive integers (0 to 65 535) in single steps, while the eight-bit microprocessor can only handle 2^8 integers (0 to 255) at a time. The address bus must be wide enough to cope with the amount of memory needed, and 24 bits (equivalent to more than 16 million memory elements) is not uncommon.

The program contained in part of the memory consists of a sequence of instructions that are located by the program counter and interpreted by the *instruction decoder*. This decoder breaks the instruction down into yet simpler logical operations that are performed by the *arithmetic logic unit (ALU)*. The ALU then carries out tasks that include arithmetic (for example, addition and subtraction), logic (for example, AND or OR — see p. 19), and comparison (for example, one number equal to, greater than, or less than another). More complex manipulations can be assembled from combinations of the simpler ones. For example, multiplication can be carried out as a series of additions, and the comparison of 'greater than or equal to' can be constructed out of the two simple inequality tests and the logical OR function. The *accumulator* is an internal memory, or *register*, used by the ALU for the local storage of data. Most microprocessors have other internal registers used for data storage and in connection with the addressing of external memory, but they are not indicated in the figure.

Pulses from a clock trigger each of the individual electronic steps that make up the operations described, including transfer of data to and from memory, decoding, and the internal arithmetic and logic. The clock may be a quartz-controlled oscillator on the chip itself, or may be external to it. Typical frequencies can be up to 40 MHz or more for modern microprocessors. Most internal operations take only a few clock cycles, so that millions of arithmetic or logical steps can be performed in each second.

An important feature of computer programs is that they sometimes need to make a *jump* to an address other than the next one in the current sequence, and particular power is imparted by the possibility of the *conditional jump* that is carried out if the results of one or more comparisons meet certain requirements. In this way, the action taken by the computer, and the program it subsequently executes, can be modified by the conditions encountered. This capability not only allows numerical calculation of enormous complexity, but also permits the adaptive control strategies that are the subject of the next section.

6.3 Microprocessors in compensation and control

An interesting comparison of the different ways in which microprocessors can be used to improve instrumentation is afforded by their use in the *compensation* for temperature effects and their *control* of the temperature itself. Measurement of pH (see also pp. 48–49) provides an important example in chemistry where changes of temperature can have an effect on the value obtained. The two approaches are either to allow for the known effect of temperature or to keep the temperature constant. The first method can, of course, be applied by hand, with the raw reading being corrected according to a formula or by looking up a table of corrections. It is clearly very convenient if this *look-up table* can be stored in the memory of a computer and the computer asked to carry out the calculation of the corrections before displaying the final result. Figure 6.2 shows the essential elements of

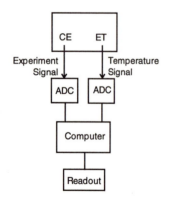

CE = chemical experiment
ET = electrical thermometer

Fig. 6.2 Experiment with temperature compensation

temperature compensation for any experiment, not necessarily the measurement of pH. Analogue electrical signals are obtained from both the experimental system and from a temperature sensor. They are both digitized for use by the computer. In the figure, two separate ADC devices are shown for clarity, although a single ADC might, in reality, be switched regularly from one task to the other. The computer now looks up the value in its table for the particular digital number that corresponds to the temperature. That is, the temperature has been converted to an address in memory. The number stored in that address is used to correct the number corresponding to the experimental quantity being measured, perhaps by simple multiplication or perhaps by using a more complex formula. In some cases, the look-up table might be stored permanently in the computer, while in others it might be generated by the computer itself for a particular set of conditions.

Figure 6.3 illustrates a temperature-controlled experiment. The control can, of course, be achieved by conventional analogue techniques such as that represented by Fig. 6.3(a). This is the most usual type of thermostatic heater. As described on p. 51, heat losses to the surroundings provide the cooling that balances energy input from the heater, with the error signal from the sensor being amplified to provide the necessary current. The success in rapidly achieving the required temperature, and maintaining it, without overshoot demands choice of a suitable control law, with proper damping, as discussed on p. 60. It may be difficult to approach the ideal control law with analogue electronics. The computer-controlled device can have the control law tailored by the software program in memory. Figure 6.3(b) shows the barest outline. The input from the sensor to the computer is digitized as in Fig. 6.2 by an ADC. The deviation from the desired temperature is now a quantity in numerical form within the computer, and can be altered mathematically in any fashion desired. The resulting number is then converted back to an analogue heating current by being interpreted by a DAC, and amplified. The ease of performing mathematical operations on the numbers, compared with imposing electrical responses on the voltages, is what gives the computer system its great advantage. The PID control law (p. 60) can readily be implemented, but so can any other. The computer can sense the rate of approach to the desired temperature and, knowing the thermal characteristics of the heater, keep a high input power applied until the temperature reaches a certain point short of the target. Since it is impossible ever to obtain *exact* temperature control, a next step in using the computer might be to apply temperature compensation corrections for the remaining small inevitable fluctuations in temperature of the experiment.

The diode-array spectrometer discussed on p. 78 and illustrated in Fig. 5.24 provides a rather different example of the use of the computer in instrumentation, since all four functions listed on p. 84, including control, can be harnessed. Figure 6.4 illustrates how a computer may be employed in *data acquisition* from such a spectrometer. An electronic switch (probably with sample-and-hold circuitry) can be instructed to connect each diode element in turn to the ADC interface with the computer, and the computer can then store

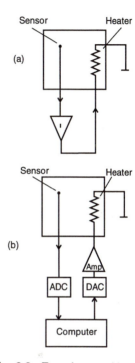

Fig. 6.3 Experiment with temperature control:
(a) analogue; (b) computer controlled

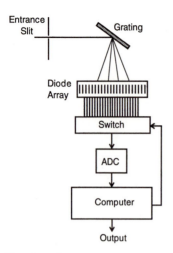

Fig. 6.4 Diode-array spectrometer (see Fig. 5.24) with computer acquisition of data

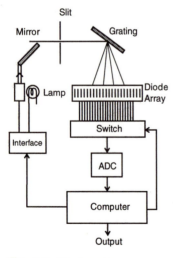

Fig. 6.5 Diode-array spectrometer of Fig. 6.4 modified to include automated calibration

each digitized voltage in an appropriate memory location. The multiplex advantage is not necessarily lost in this sequential scanning, because each individual diode can be accumulating information at its sample-and-hold circuit throughout the experiment, and the read-out may occupy only a short period compared with the sampling period. The computer can also obtain time-integrated or time-averaged signals of its own by electrically scanning the diodes repeatedly, and adding the results to those already in memory for the appropriate channel. This operation is one of *data processing*.

One of the problems with the diode arrays is that each diode has a slightly different response to the same intensity of light, and there is also likely to be a more marked difference in response to wavelength of radiation. Each element thus produces an output signal that depends partly on its own characteristics and partly on where in the image plane of the spectrum it lies. A solution that exploits *data processing* again uses the compensation methods described earlier. A look-up table can be generated by a preliminary calibration procedure, so that the data from the computer can be normalized mathematically before output. Further sophistication can be achieved by using the computer for *experiment management*. Figure 6.5 shows the computer able to turn a lamp on and push up a mirror to reflect its light into the spectrometer slit as well as perform the other tasks that we have already described. At suitable intervals, the computer can ask for this lamp and mirror combination to be activated, so that the diode array sees the light from a source whose characteristics can be well defined. The signals from the diodes can now be used to provide a calibration graph for the array. In reality, the calibration data would be used to reconstruct a new look-up table to provide updated compensation. This rewriting of part of the software can be regarded as a subtle form of *feedback control*.

The use of compensation information that is derived from the experiment itself is the first step in constructing so-called 'intelligent' or 'smart' instruments that adapt their strategy according to the conditions encountered instead of following a fixed routine. Such *adaptive* or *optimizing control systems* are having an important impact on the capabilities of scientific instruments. A simple example, again taken from the field of spectroscopy, might involve a device designed to examine very weak fluorescence, this time with a monochromator–photomultiplier combination. The problem with such a scanning instrument used conventionally is not only the signal wasted outside the spectral band being examined (see pp. 77–78), but also that even more potential signal acquisition time is wasted if there is only a limited number of relatively sharp spectral features. Under adaptive computer control the scan can be set to be much more rapid than would normally be acceptable for the required signal-to-noise ratio, and the computer can issue instructions to slow the scan down only if intensities above some predetermined level are found. The available experimental time can then be used to improve the precision of the intensity measurement of the spectral features that are present rather than in improving the precision with which it is known that no light is falling on the photomultiplier.

Further reading

I have found the first three books particularly helpful during the preparation of the present text. They cover virtually all the topics discussed here, often in considerably greater depth; they also provide an introduction to the elements of circuit design. The other books present other approaches to instrumentation and to electronics; they are listed alphabetically.

P. Horowitz and W. Hill. *The art of electronics* (2nd edn). Cambridge University Press, Cambridge, 1989.

B. K. Jones. *Electronics for experimentation and research*. Prentice-Hall, Englewood Cliffs, NJ, 1986.

H. V. Malmstadt, C. G. Enke, and S. R. Crouch. *Electronics and instrumentation for scientists*. Benjamin/Cummings, Reading, MA, 1981.

J. J. Brophy. *Basic electronics for scientists*. McGraw-Hill, New York, 1990.

D. Buchla and W. C. McLachlan. *Applied electronic instrumentation and measurement*. Maxwell Macmillan International, New York, 1992.

A. De Sa. *Principles of electronic instrumentation* (2nd edn). Edward Arnold, London, 1990.

P. R. Griffiths. *Transform techniques in chemistry*. Plenum Press, New York, 1978.

R. L. Havill and A. K. Walton. *Elements of electronics for physical scientists* (2nd edn). Macmillan, London, 1985.

L. D. Jones and A. F. Chin. *Electronic instruments and measurements*. Prentice-Hall, Englewood Cliffs, NJ, 1991.

G. Long. *Real applications of electronic sensors*. Macmillan Education, Basingstoke, 1989.

N.C. Morris. *Electronics: practical applications and design*. Edward Arnold, London, 1989.

T.H. O'Dell. *Circuits for electronic instrumentation*. Cambridge University Press, Cambridge, 1991.

P.P.L. Regtien. *Instrumentation electronics*. Prentice-Hall, London, 1992.

T.H. Wilmshurst. *Signal recovery from noise in electronic instrumentation*. Hilger, Bristol, 1990.

Index